Science & Religion

"An Astronomical Explanation of the Fatima Miracle"
> Comet Fatima <

A scientific investigation of what

happened in Fatima in 1917

Real Science with social implications

It was not a hoax or mass hallucination

We explain a mystery and uncover another

With 84 diagrams and illustrations

Science Touches the Divine

Ignacio R. Ferrín V., Ph. D.

© 2021, Ignacio R. Ferrín V.
all rights reserved

No part of this book can be reproduced without permission, in any form, and by any means.

ISBN: 978-958-49-1920-5
ISBN: 978-958-49-1921-2
ISBN: 979-873-99-0264-1

You may contact the author at

nextimpact2022@gmail.com

Feedback is always welcome.

Version 1: 2021-05-13, exactly 104 years after the first apparition.
Version 2, 2021-10-03.

Cover: Fatima comet approaching planet Earth, and witnesses at the Cova da Iria place, watching the Miracle of the Sun in 1917, without eye protection. (Credit: NASA and Judah Bento Ruah).

This work contains research that uses data or services provided by:

- The International Astronomical Union's Minor Planet Center,
- NASA Jet Propulsion Laboratory, JPL, California Institute of Technology, Small Body Data Browser,
- IAU Meteor Showers Database.
- National Observatory of Venezuela, CIDA.

Without these resources, this book would not have been possible.

I dedicate this work to
Alice,
Celeste,
and Arsenia.

INDEX

Preface: Why I wrote this book 7

Chapters

1- The Scientific Method 11
2- Origins 17
3- Historical events 33
4- Image Interpretation 43
5- Events Previous to the Miracle 53
6- Comet Fatima: Analysis I 89
7- Comet Fatima: Analysis II 137

Index of citations and collaborators 151
Glossary 153
References 156
About the author 162
Acknowledgements 164

Preface
- Why I wrote this book

> *"I believe in evidence,*
> *I believe in observations,*
> *measurement and reasoning,*
> *confirmed by independent observers.*
> *I believe* **anything**,
> *no matter how wild and ridiculous,*
> *if there is* **evidence** *for it."*
>
> *"Anti-intellectualism has been a constant*
> *in political and cultural life,*
> *nurtured by the false notion*
> *that democracy means*
> *that "your ignorance is just as good*
> *as my knowledge"."*
>
> Isaac Asimov

- Why I wrote this book

I spent several years writing a book with the title *"Next Asteroid Impact,"* where I give new information on the asteroid and comet threat on our planet. While I was doing that, I had to do a lot of research on the internet, gathering meaningful information on our solar system's minor bodies. As a result, I came across several notable events that have affected our planet in the past: the fall of a bolide over Madrid that broke windows in 1896, the fall of an asteroid over Tunguska in 1908 that leveled 2500 square kilometers of forest, and the event of Chelyabinsk in 2013, a Russian city, that caused 1700 wounded.

I found many events, some small, some more significant, but one stood out as very significant. It happened over a tiny region of Portugal, Cova da Iria. What appeared was strange and difficult for the witnesses to report. The observers described the Sun as "dancing," that it had "changed colors," and that it had "fallen" over the people.

My book was progressing along, and as a byproduct, new circumstantial evidence having to do with Fatima was being uncovered and accumulated, until one day I realized that I had a scientific and astronomical explanation of what had happened in 1917 in Portugal! I couldn't believe it!

I abandoned the other Chapters of the book and began to concentrate on the Fatima Chapter, until the information grew to such an extent that a separate book had to be written to include all the material gathered. This is the book you have in your hands now.

As I worked through the various chapters and through the many revisions, I found myself focusing on the main chapter where the event at Fatima is empirically and scientifically explained. I worked tirelessly to complete that chapter. After finishing my revisions, I chanced to look up at my calendar and the date was May 13, 2020, the date of the first reported apparition of the Lady to the three children at the Cova in Fatima. I could not believe it! This was the 103rd anniversary of the first of the purported apparitions. The probability of this happening is 1 in 365 or .003. Small but not that small! But that was not all.

When I was ten years old, I was attending a school of the Marist Brothers, and on May 13th of every year, they took us all to a nearby chapel where we chanted this song:

> On May 13th
> the Virgin Mary
> descended from Heaven
> to Cova da Iria

At that time, I was a child, and I could not have guessed that I would research and decipher the event many decades later.

What my research discovered is that a small comet falling over Fatima could explain all the atmospheric phenomena observed. So it was neither a hoax nor a mass hallucination. It was a real physical event. And it was possible for me to decipher this event thanks to my training of decades researching comets. But having solved this mystery, another one came up, a puzzle that has no solution in the present scientific realm.

I am a scientist and an educator. As a scientist, I have to search for the truth, and follow ALL the evidence wherever it leads, even if it leads to esoteric or unfathomable places. As an educator, I have to spread that scientific truth. In this book, you will find all the evidence I have gathered, presented for your evaluation and criticism. Respectful criticism is the best way to perfect truth. I have to ask your forgiveness if a chapter becomes somewhat heavy in scientific arguments. But they have to be made for the sake of building the case. In that regard, there is not one single equation in this book.

These are the reasons why I wrote this book.

I hope you will enjoy this story as much as I enjoyed researching and writing it.

Figure 1. Galileo Galilei (1564-1642), painted by Domenico Tintoretto. The word in Latin "MATHUS" means Mathematician. Inset: Two of his telescopes. The upper one has a lens with a diameter of less than 2 cm, ~0.8 inch (Credit: Public Domain).

Chapter 1
- The Scientific Method (SM)

In this Chapter we are going to do a brief review of the Scientific Method (SM). The SM is the most potent methodology to understand Nature, so it pays to expend some time to understand fundamental concepts like *"fact"* and *"hypothesis."* We will need these concepts to carry out our investigation of the Fatima event.

1- Introduction

Science is a body of knowledge that is open to everybody with common sense. Its methodology is the Scientific Method (SM). Galileo Galilei (1564-1642), born in Pisa, Italy, was the first to develop the Method. We can put a date on its inception: November 30th, 1609. On that night, Galileo pointed his telescope at the Moon and thus initiated the Era of observing Nature systematically.

Galileo (Figure 1) did not invent the telescope but was the first to put it to good use, observing the cosmos. He discovered amazing things that were unknown at the time: craters on the Moon (Figure 3), satellites of Jupiter, the phases of Venus, spots on the Sun and "ears" on planet Saturn. His observations led him to concluded that the Sun was at the center of our solar system.

After many generations, through a sequence of students, his knowledge reached us, and we will pass it on to our students (Figure 4). We owe Galileo not only the SM but many of our advances in our civilization, from airplanes to medicines, computers, bridges, trains, electricity, vaccines, the transistor, and much more.

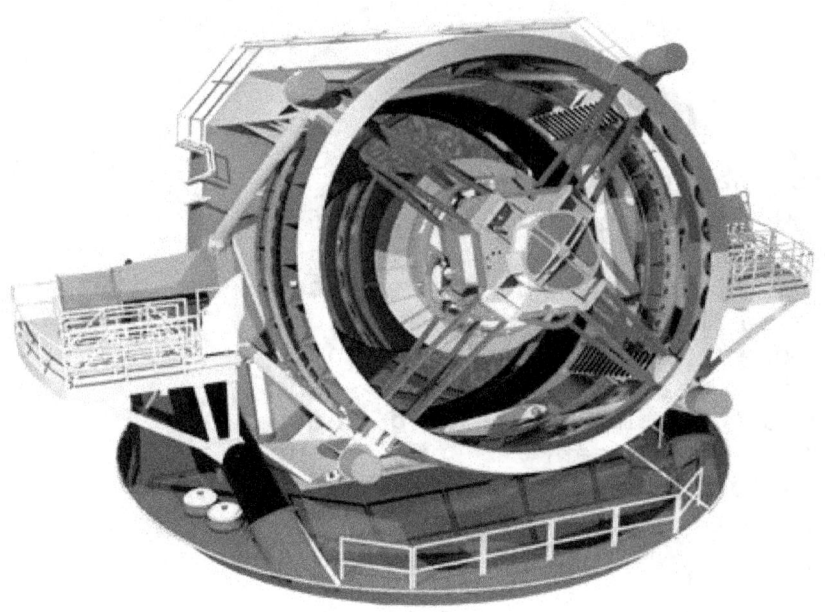

2021

Figure 2. The new Vera Rubin Telescope being built in Chile, will be operational in 2023, and has a mirror 6.5 meters (21 feet) in diameter. Compare with Galileo's telescope in Figure 1 in 1609. (Credit: LSST Project Office).

<> <> <> <> <>

Figure 3.
A drawing of the
Moon by Galileo
Using his new
telescope.
He was the first
to recognize
craters on the Moon.
(Credit: Public
Domain).

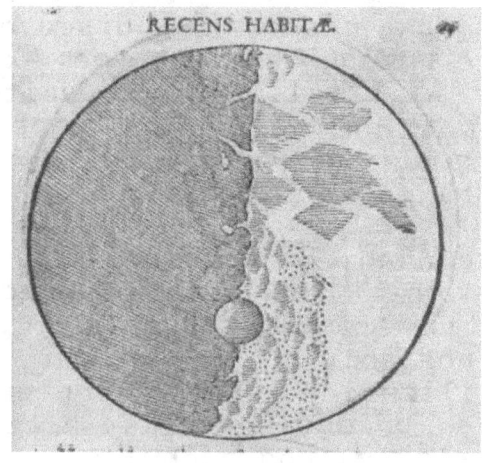

The Scientific Method encompasses a set of rules and procedures, which have to be followed to assure the correct interpretation of Nature. Here we will only deal with the most significant rules.

1- Rules of the Scientific Method

We will consider only those rules that we will use in this investigation. The first and most important rule is this one:

- Rule #1

"I think that in the discussion of natural problems, we ought to begin not with the scriptures, but with experiments and demonstrations."

Galileo says that knowledge comes from Nature, not from old books and texts. And he adds that experiments and demonstrations are the correct way to ask questions about Nature. That is very clear to us today, but it wasn't clear during the epoch of Galileo.

- Rule #2

"The properties of Nature are coded in Laws that are unchanging, universal and inviolable."

If we want to understand Nature, we have to discover the Laws that govern it. These Laws can't be changed and are permanent and universal; that is, they apply to the whole Universe. For example, the Law of Gravitation of Newton extends to the far edge of the Universe. And the Theorem of Pythagoras can be applied here or on the star Alpha Centauri, and the result will be the same.

- Rule #3

Lord Kelvin once wrote:

"If we cannot express it in numbers, it is not Science; it is opinion."

Science does not work with opinions. It works with facts that can be demonstrated numerically. That is why love and hate are not scientific terms, because they cannot be expressed in numbers.

- Rule #4

> *"Science works with "facts." A fact is something that has happened, without doubt, that can be proved scientifically, and it is there for everybody to see. Facts do not belong to anybody in particular. They belong to the whole of humanity."*

One of the essential rules before an investigation can be launched, is that we have to agree on the facts. For example, it is a fact that the Sun rose today. Anybody can look through the window and see that the Sun is there. And it is there for everybody to see. A scientist may have the results of an experiment in his laboratory, a fact, but if you go to his lab, you will be able to see it too.

- Rule #5

> *"All research projects have hypotheses."*

A hypothesis is an idea, a conjecture, a possible explanation of reality that has not yet been demonstrated as valid. A research project has to have at least one but possibly several hypotheses. The objective of the investigation is to find their validity. On some occasions hypotheses are formulated as a question: "Does this medicine cure cancer?" is a hypothesis.

It is the work of the scientist to declare hypotheses as TRUE or FALSE.

- Rule #6

> *"In modern times, the results of investigations are published in international journals that are refereed."*

Figure 4. When Galileo became blind, Jose de Calasanz, founder of the Piarist Religious Order and friend of Galileo, asked another priest, Clemente Settimi, to be in service of Galileo, attending him and being his secretary. In this way he could pass his knowledge on to his disciples, and on to us. In this painting by Cesar Cantagalli (1870), we see the young Clemente listening to and writing down the teachings of the old and blind Galileo. Calasanz never abandon him, and continued supporting Galileo until his death. (Credit: Public Domain).

Refereed means that the submitted work will be evaluated by one or several judges (referees), who will oversee that the SM rules are being followed.

This author abides by the rules of the Scientific Method.

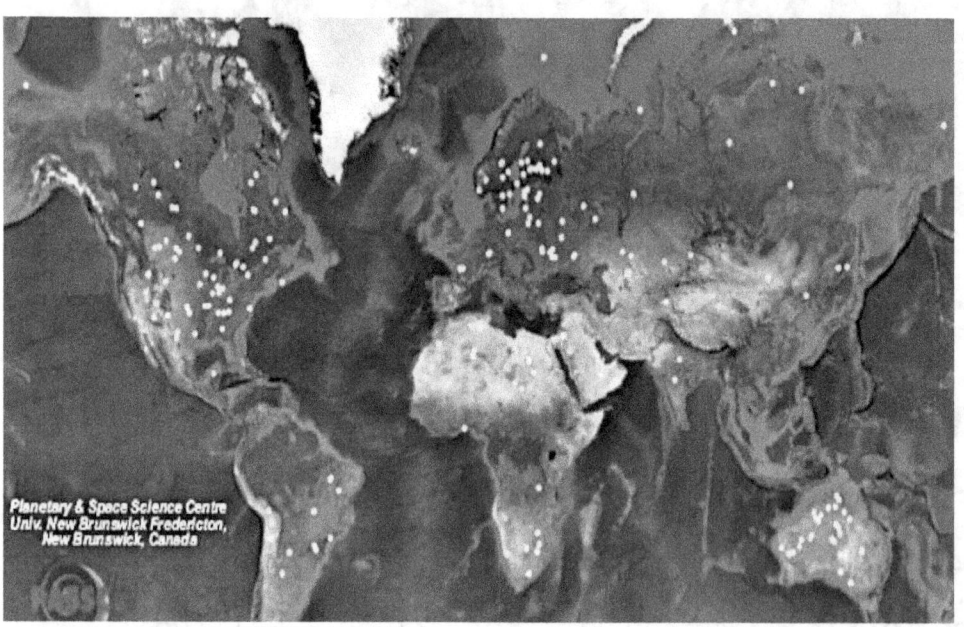

Figure 5. World impacts of asteroids. Notice that unpopulated areas (Siberia, Amazonia, and Central Africa) have few discovered impacts. Impacts seem to have been discovered in densely populated areas (USA, Europe, and Australia). So there may be many undiscovered craters in unpopulated areas. (Credit: University of New Brownswick, Canada).

CHAPTER 2
- Origins

In this Chapter we will investigate the origins of the fall of comets and meteors on our planet, setting the scene to comprehend Fatima's event. Please refer to the Glossary at the end of the book if you are not familiar with some terms.

1- Introduction

This detective story has been developing for millions of years (Figure 5), with a high point 65 million years ago with the dinosaurs' extinction (Figure 6). It continued up to the present with a total of about ~300 impacts on our planet[1,3]. The difference with Moon impacts is that the Moon does not have an atmosphere, so impacts are preserved untouched. But on our planet the wind, rain and vegetation tend to erase any trace of shallow craters. If we look at Figure 5, we see that unpopulated areas (the Amazon, Siberia and central Africa) have few or no craters. In contrast, most known craters lie in populated areas (USA, Europe and Australia). Thus there is no doubt that there are undiscovered craters in unpopulated areas.

This was the epoch of "heavy bombardment" that ended about 4 billion years ago, believed to represent gigantic collisions due to the residual matter left from the solar system formation.

2- Extinction of the dinosaurs

This story's first notable date was the disappearance of the dinosaurs during the Cretaceous-Tertiary geologic age (Figure 6). This took place about 65 million years ago. Research points to the crater Chicxulub as the culprit (Figure 7).

Figure 6. It is believed that the dinosaurs went extinct 65 million years ago when an asteroid greater than 10 km in diameter impacted the Yucatan peninsula in Mexico. The smoke produced by the fires and the dust raised by the impact covered the atmosphere with a layer so opaque that the solar light did not reach the surface, extinguishing all vegetation and food. (Credit: Wikimedia Creative Commons, public domain).

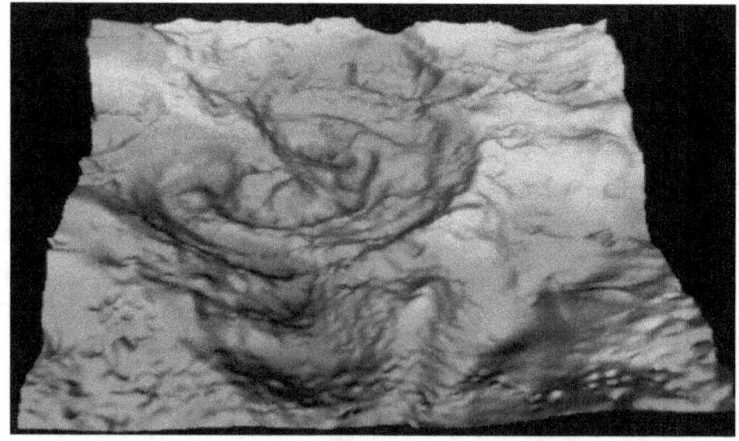

Figure 7. Computer-generated gravity map of the Chicxulub crater. The main structure is underwater. The diameter is about 180 km. Notice the channel area in the lower part of the picture, suggesting that the asteroid entered at a low impact angle. The outer ring is about 300 km in diameter. Colors indicate depth, with blue the deepest. (Wikimedia Commons, public domain).

Chicxulub is located in the peninsula of Yucatan, Mexico. The crater is enormous, with a diameter of about 180 km. It has been estimated that the meteorite size that impacted was about 10 km or more, and the energy released equivalent to about 3.000 bombs of Hiroshima. It was the researcher Luis W. Alvarez and his son Walter who suggested the idea for the extinction[22,23].

3- Fall of meteorites

Meteorites (rocks from space) have fallen over our planet since its formation. Thousands of meteorites have been collected throughout history. In the beginning, people thought that meteorites were terrestrial rocks that had been struck by lightning. Others thought that they came from active volcanos on the Moon. But most people, even scientists, had difficulty in believing that they could come from space.

Scientific skepticism continued into the eighteenth century. On December 14th, 1807, a fireball exploded over Weston, Connecticut, and dropped several stones over the town. When Thomas Jefferson then 3rd President of the United States heard about this fall, he supposedly said: *"I would more easily believe that two Yankee professors would lie, than stones would fall from the heavens."*

In 1985 the United States army deployed a series of military satellites to detect possible atomic explosions to monitor nuclear treaty enforcement. Consequently, military analysts were stunned to find very intense bursts in the atmosphere that had nothing to do with atomic artifacts.

When the scientists realized they had detected atmospheric explosions, they immediately knew that meteorites had caused them. After negotiations with the military, they agreed to release the data for scientific studies, and the result was incredible. Since 1988 more than 850 events have been registered and studied in detail.

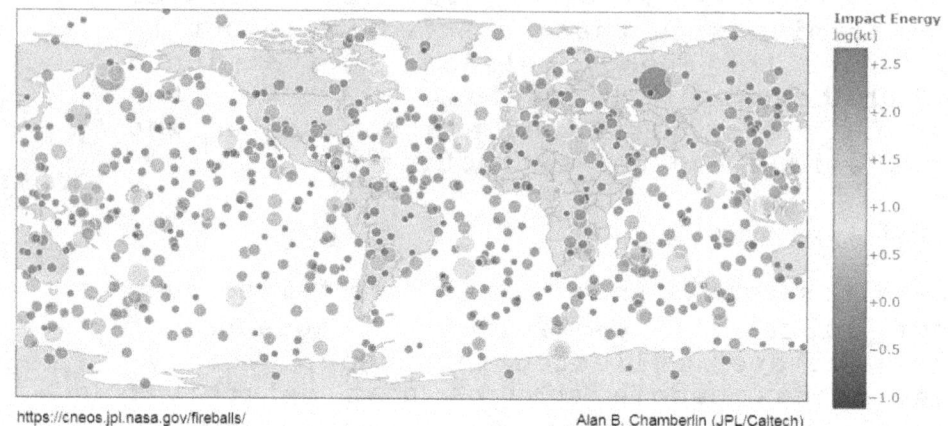

Figure 8. The CNEOS database has registered more than 850 explosions in the atmosphere believed to be caused by impacting asteroids. When they burn in the atmosphere, they produce meteorites on the ground. These are only the most intense events. The equipment does not register the faint events. (Credit: CNEOS, Alan B. Chamberlin (JPL/Caltech).

Figure 8 shows the CNEOS (Center for Near-Earth Object Studies) web page with 850 impacts plotted rather uniformly over the planet. The red dot in the right hand side is the Chelyabinsk event.

3- The discovery of the asteroids

Up to 1801, the planets known were Mercury, Venus, Mars, Jupiter, Saturn and Uranus, discovered by William Herschel in 1781. Neptune was found much later, in 1846, and Pluto in 1930. Astronomers had realized that there was a gap between Mars and Jupiter with no planet. Some astronomers suggested that a planet had to exist, and searches were underway to find it.

On January 1 of 1801, Gioacchino Giuseppe Maria Ubaldo Nicolo Piazzi, an Italian astronomer

Figure 9. Giuseppe Piazzi, the discoverer of the first minor planet 1 Ceres, on the night of January 1, 1801, while he was working on a new star catalog. (Credit: F. Bordiga, Smithsonian Institute Library. Public Domain).

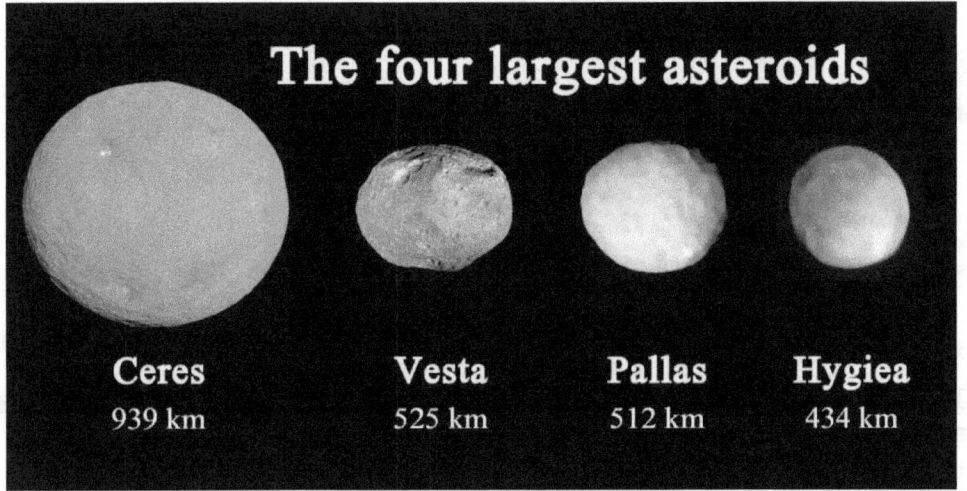

Figure 10. The four largest asteroids of the Main Belt, Ceres, Vesta, Pallas, and Hygiea, are known as "The Big Four." (Credit: NASA/JPL-Caltech/UCLA/ MPS/DLR/IDA, public domain).

<> <> <> <> <>

(Figure 9), worked on a new star catalog. At about 8 pm, he cataloged a star in the "shoulder" of the constellation Taurus. Piazzi registered its position and, as usual, measured the star again the next night. Surprisingly it had moved. He thought it was a mistake at first, but he concluded that it might be a new comet after four days.

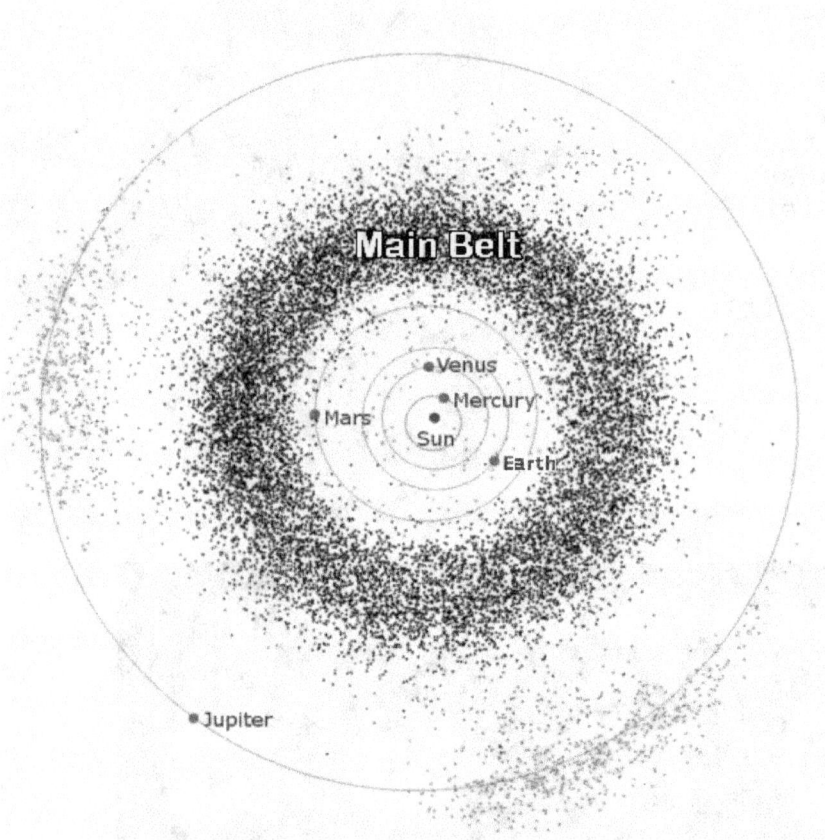

Figure 11. The Main Belt of Asteroids is located between the orbits of Mars and Jupiter. More than 1.000.000 objects are orbiting around the Sun in this area, including the one that fell on Fatima. I present this image as a negative because positive photos with black background tend to erase fine detail, and there is a lot of fine detail in this image. A large fraction of these objects tends to cross the Earth's orbit dangerously. They are called PHAS, Potentially Hazardous Asteroids. (Credit: Wikipedia Commons, public domain).

<> <> <> <> <>

On those days, the only objects in the sky that moved were planets and comets. So Piazzi's conclusion was reasonable but wrong. The discovery was announced as a comet. He sent his observations to another astronomer, Oriani, who was the first to suggest that it might be a new planet.

A new planet it was, and we know it by the name of 1 Ceres (Figure 10), the first member of the *"Main Belt of Asteroids."* In the following years, thousands of new asteroids were discovered. By mid-2020, the number of asteroids going around the Sun up to and beyond Pluto's orbit reached more than one million objects (Figure 11).

4- Comets in the Main Belt of Asteroids

By 1979 our knowledge of the Main Belt had increased enormously. Individual asteroids were studied in detail to obtain their physical and orbital properties. But there was a puzzling fact that was perturbing that knowledge. Among the thousands of asteroids known at that moment, there was an outlier, a comet under the name of 2P/Encke. It was a strange situation. All those asteroids were bare stones, but what was that single comet doing there among so many asteroids? From where did it come? How did it get there?

Since 2P/Encke was the only known comet in the Main Belt, the suspicion was that it had come from outside the Belt, probably from the region of Jupiter. Numerous calculations were made to model the transition in space, but all ran into difficulties. The time to make the transfer was millions of years. The probability of transfer was minimal, so it did not look like astronomers had found an explanation. But the paradigm of the Main Belt exclusively composed of bare rock was about to change.

On November 19th, 1949, astronomers A. G. Wilson and R. G. Harrington discovered a comet in the first Palomar Observatory Sky Survey (POSS) using the Schmidt camera. It exhibited cometary activity in two images, one of which is reproduced in Figure 12. From there on, we lost it.

On July 24th, 1979, the Siding Spring Observatory in Australia discovered asteroid 7968. Later on, in 1996, Eric W. Elst of the Royal Observatory at Uccle reported that in images of

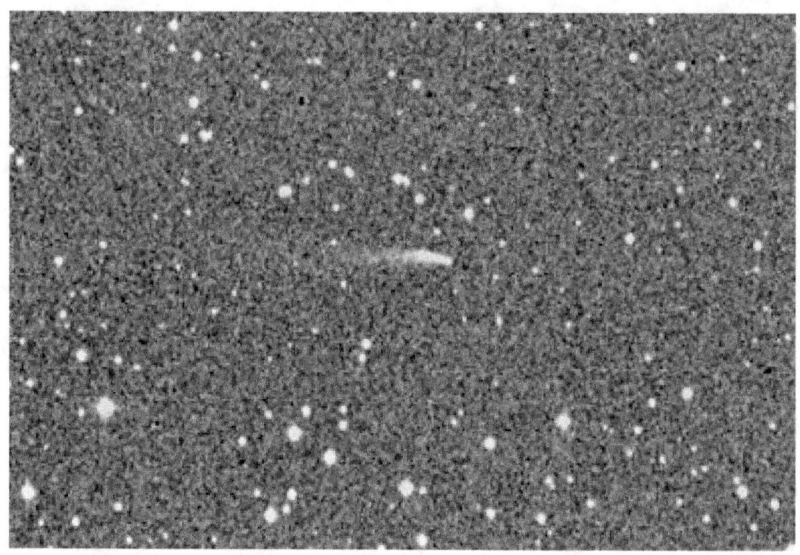

Figure 12. Minor Planet 4015 on a blue 12-minute photographic plate, obtained on November 19, 1949, with the 48-inch Schmidt telescope of the Palomar Observatory. The image was enhanced at the European Southern Observatory (ESO) photographic lab in Garching to show the tail better. If an asteroid exhibits a tail, it implies that it is not an asteroid but a comet. As a result, the asteroid was renamed comet 107P/Wilson-Harrington. (Credit: ESO and Palomar Observatory, Wikipedia Commons License (CC BY-4.0).

7968 taken by Guido Pizarro, they had detected the continuous presence of a tail. This observation suggested that it was not an asteroid as previously thought, but a comet, and thus was renamed 133P/Elst-Pizarro (Figure 14).

Just four months later, in the same year, another asteroid was discovered at Palomar Mountain Observatory in California by the astronomer Eleanor Helin on November 15, 1979. First, it received an asteroid number, 1979 VA, and a few years later, a permanent number, 4015.

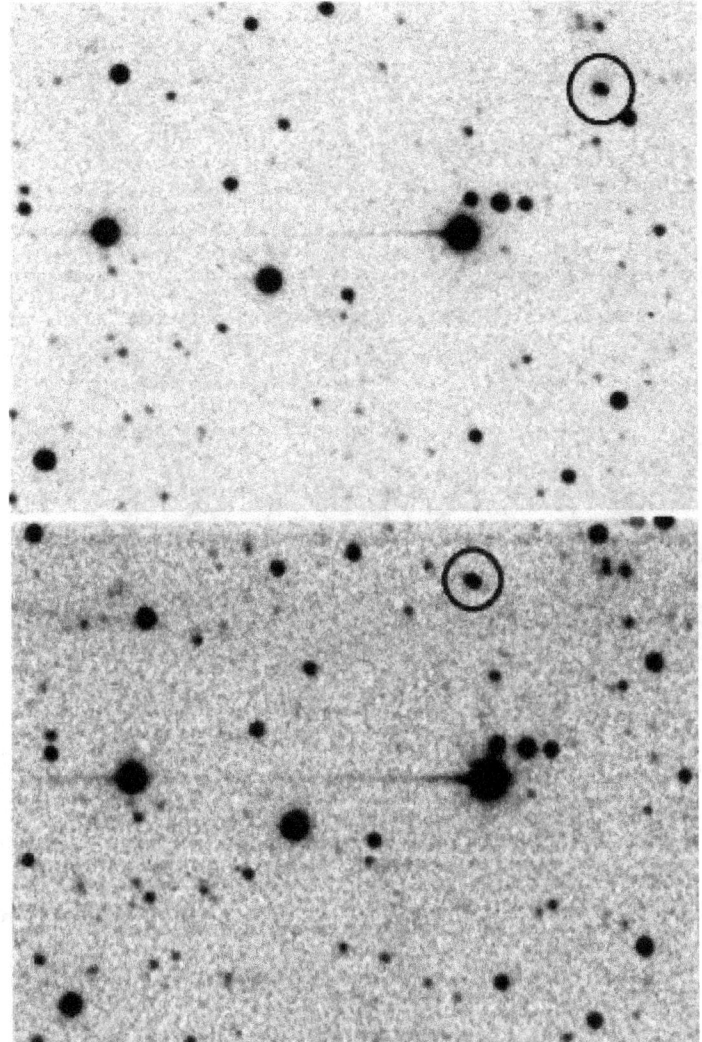

Figure 13. Comet 107P/Wilson-Harrington was imaged by this author using the 1.0 meter Schmidt telescope of Venezuela's National Observatory (Figure 50). At the time, it did not show any cometary activity. (IF24).

<> <> <> <> <>

    Brian G. Marsden, the ex-Director of the Minor Planet Center, was an expert in orbit calculation, and he soon realized that the object observed in 1949 had the same orbit as the asteroid discovered by Helin. Marsden concluded that the comet was inactive, perhaps because it was ancient.

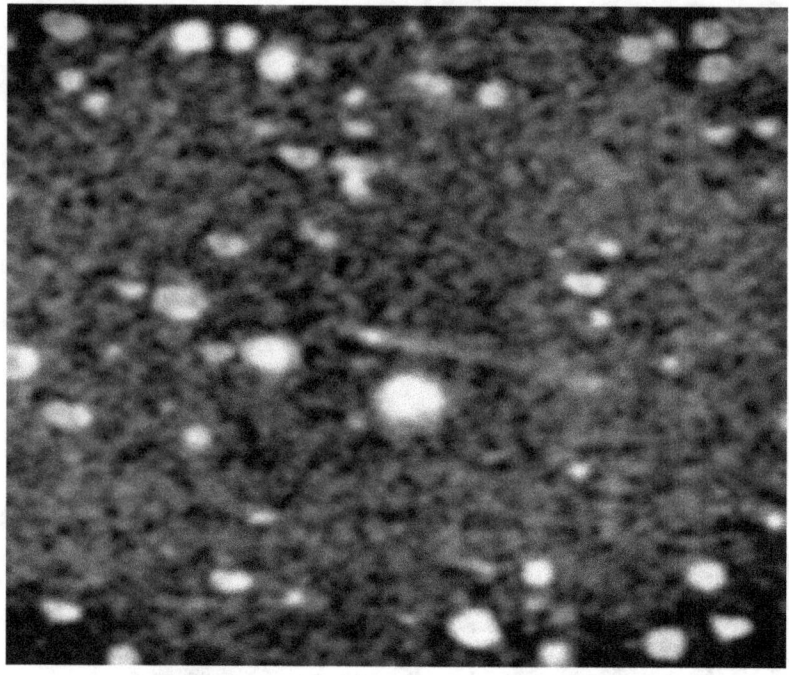

Figure 14. This image of asteroid 7968 was taken by this author with the 1 meter Schmidt telescope of the National Observatory of Venezuela (Figure 50), at 3600 meters of altitude. It shows the asteroid exhibiting a tail. Thus, it cannot be an asteroid. It has to be a comet, the third one discovered in the Main Belt of Asteroids. Accordingly, it was renamed 133P/Elst-Pizarro. (IF[45]).

Thus by 1996, there were three comets known in the main Belt, 2P/Encke (Figure 46), 107P/Wilson-Harrington (Figure 13), and 133P/Elst-Pizarro (Figure 14), the last two initially discovered as asteroids!

A popular saying claims that *"if there is one, there could be many."* By 2020, the number of active comets in the Main Belt was more than 20 by some researchers count, but in our databases, we have more than a hundred[24-28], and the number continues increasing.

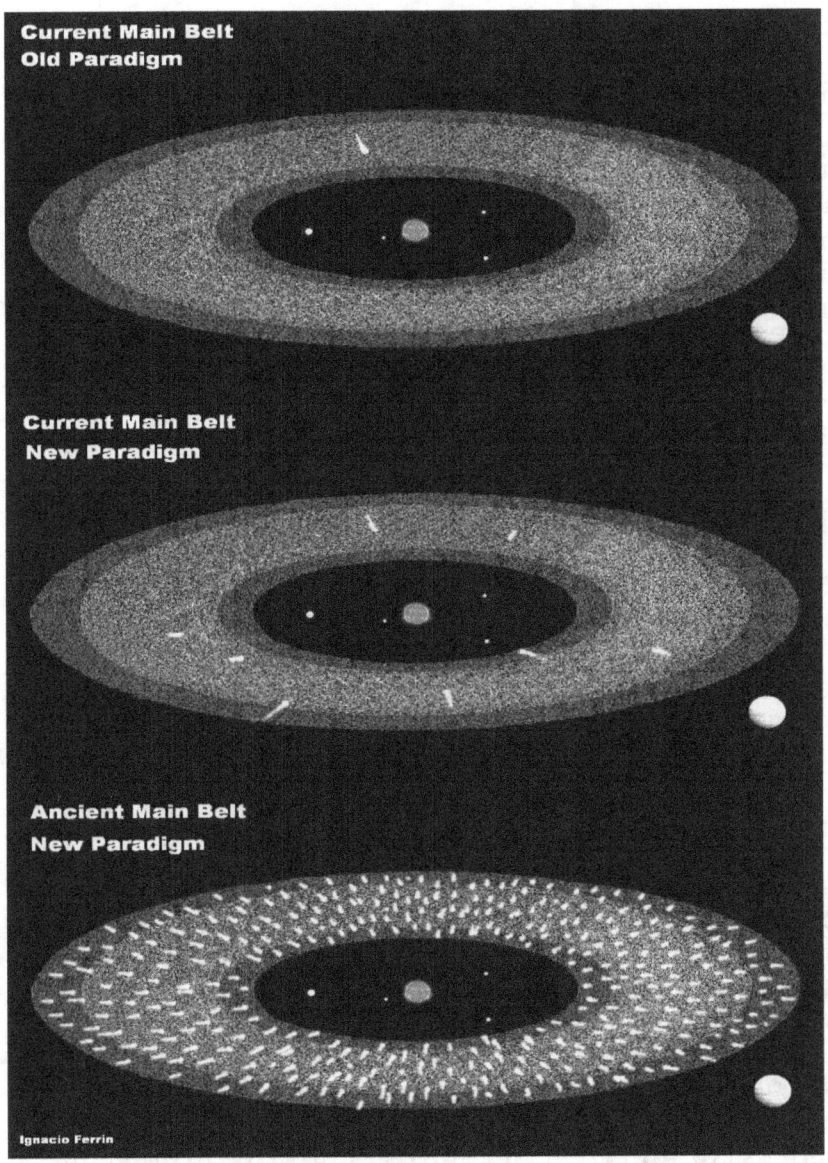

Figure 15. The Main Belt of Asteroids at three different epochs. Above: Up to 1979, the Belt was a bare rock field with a puzzling one single comet inside, 2P/Encke. Middle: By 2020, there are more than 100 objects inside the Belt exhibiting cometary activity. Bottom: Since the comets age, if they are barely active in the present, there had to be thousands of objects vigorously active in ancient times, thus creating a second Milky Way in the sky. (IF).

Comets have the property of aging, as we have seen with 107P/Wilson-Harrington (Figure 13), which was confirmed as being very old. Thus, if we found around 100 comets inside the Main Belt, there ought to be thousands in ancient times. Therefore the paradigm might have changed (Figure 15). **The Main Belt of Asteroids might not be a field of bare rocks but a graveyard of comets.**

At the beginning of this Section, we mentioned that comet 2P/Encke was an odd comet because it was the only one in the Belt up to 1979. However, if there are thousands of them, then its status has changed. The question to ask was: *"Where did it come from?"* Now we know that it was the wrong question. The right question to ask is: *"Was it there all the time?"* Yes. It came from nowhere. It had been there all the time. So the question of the origin of comet 2P/Encke has been resolved. But that is not all. 2P still has additional surprises.

5- The odd comet 2P/Encke

Comet 2P/Encke (Figure 16) is the oldest periodic comet on record after 1P/Halley. Pierre Mechain discovered it in 1786. Since it has a very short orbital period, it has returned more than 70 times to our neighborhood. But one odd feature of this object is that it has been associated with 30 meteor showers (Figure 18). 30!

Comets lose mass because of the Sun's heat in the form of gas and dust particles. The dust particles are stored in filaments that follow the comet, and when our planet intercepts these filaments, they create meteor showers, and there are more than 800 of them. Each comet has its associated shower. One comet, one meteor shower. But comet 2P/Encke has 30! How can this be possible?

W. F. Denning[29], in 1928, proposed the idea that comet 2P had 13 separate sub-streams in a meteor complex. This can be explained if comet 2P/Encke had suffered a series of disintegrations

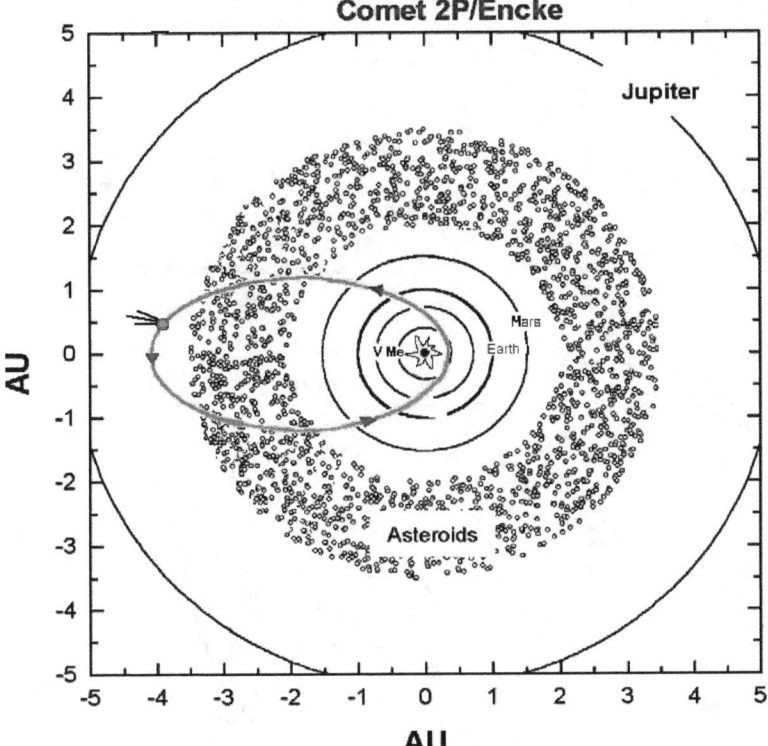

Figure 16. The orbit of comet 2P/Encke, seen from the North Pole, passes through all the Main Belt but does not reach planet Jupiter. Thus it is difficult to explain that it had come from its neighborhood. The wrong question to ask was: *"Where did it come from?"* versus the right question to ask, *"Was it there all the time?"* (IF).

in smaller objects and a lot of debris. All those minor objects and debris now form what is called the Taurid Complex.

In 2010 W. Napier[30] was able to identify 19 members of the Complex, and in 2021 my colleague Vincenzo Orofino[28] of Salento University in Lecce, Italy, and I, were able to identify 89 objects shown in Figure 17.

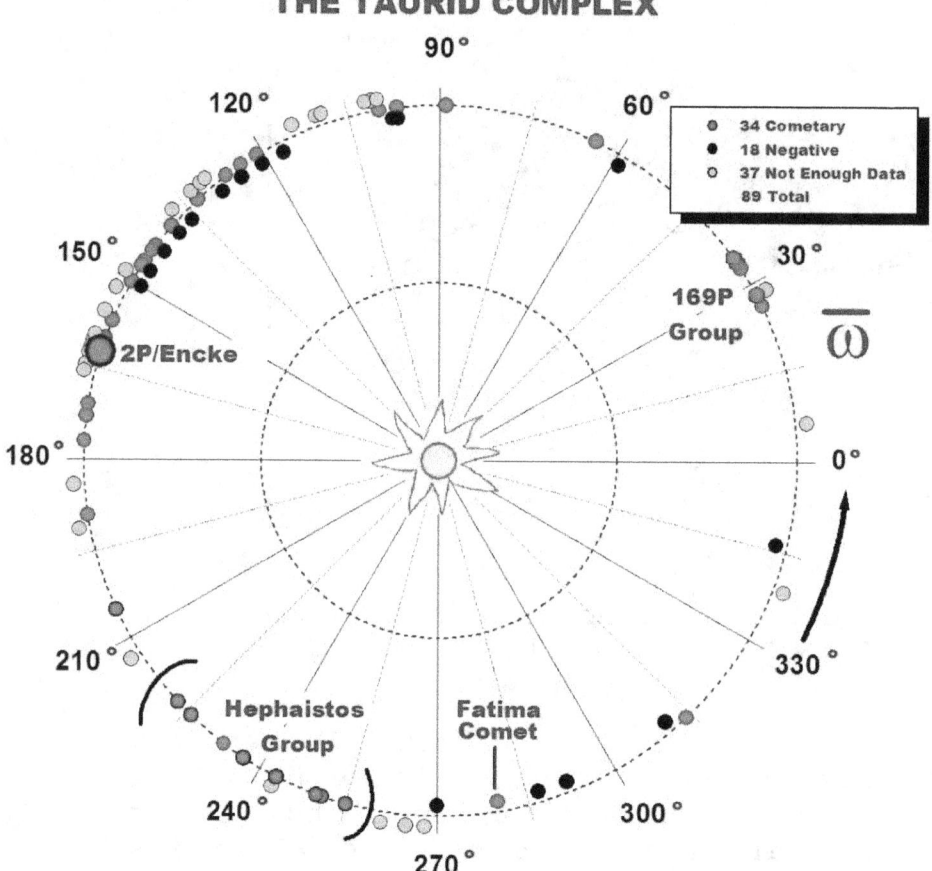

Figure 17. Comet 2P/Encke is the Taurid Complex parent, shown in this diagram in all its luster. The orbit of comet 2P/Encke is elliptical, but we show it as circular for clarity. We see that contrary to typical comets, 2P has more than 89 fragments, plus dust, pebbles and debris, accumulated along its orbit due to a cascade of historical disintegrations. The fact that comet Fatima is far from the position of comet 2P implies that this fragment was one of the first to be detached from the parent 2P. (Credit: Vincenzo Orofino and this author[28]).

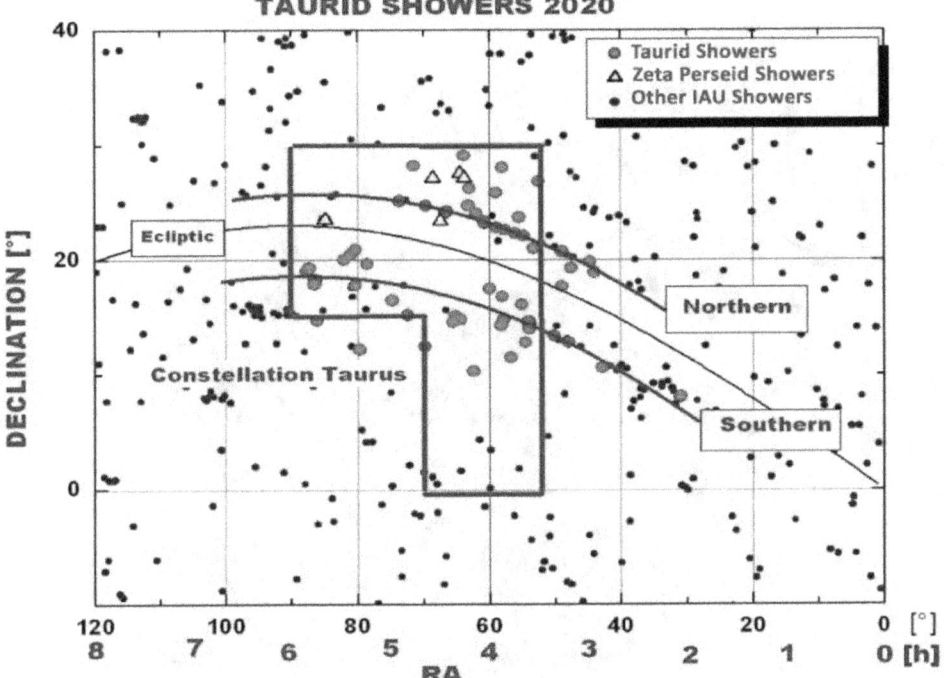

Figure 18. Comet 2P/Encke is the Taurid Complex parent. It has been associated with about 30 meteor showers, a remarkable number not seen in any other comet. The showers lie in the constellation Taurus, and there are two branches, the Northern and the Southern. (Credit: Vincenzo Orofino and this author[28]).

So what is the explanation for the 30 meteor showers associated with comet 2P/Encke (Figure 18)? The answer is that some of the 89 smaller objects associated are active comets, each of which has its associated shower.

At this time, you may be wondering why we are following the trail of comet 2P/Encke. You may have guessed. The reason is that Comet Fatima, of which we will talk in the next Chapters, may be a member of the Taurid Complex illustrated in Figure 17.

Figure 19. The Tunguska asteroid compared with the Statue of Liberty. Notice the human figure at the lower right hand side of the image. (Credit: Alvaro Gracia Montoya, a motion graphics and 3D designer of Birmingham, UK, creator of MetaBallStudios).

CHAPTER 3
- Historical Events

In this Chapter we will review some of the most important events that our planet has experienced, particularly the Tunguska event of June 30th, 1908, and the Chelyabinsk event of February 15th, 2013.

1- The Tunguska Event

On the morning of June 30, 1908, a celestial body of about 60 meters in diameter (Figure 19) struck the area of Tunguska in the middle of the Siberian taiga (Figure 20), in the early morning (Figure 21), at about 7:14 am local time[33]. The Destruction Power was equivalent to 333 Hiroshima bombs (Figure 23). It leveled an area of trees of about 2500 km^2 (Figure 22). The object moved from the SE to the NW, followed by an explosion. Seismic and acoustic phenomena were registered in the Irkutsk magnetic station seismograph. The perturbation wave moved at 330 meters per second, practically identical to the sound velocity.

L. A. Kulik led the first expedition to the site in 1927[31], 19 years after the event, thanks to the U.S.S.R. Academy of Sciences support. As of 2020, more than 50 expeditions have been made. Kulik took photographs of the site in 1927 (Figures 22, 25), and aerial images were available by 1929. Although holes were drilled in search of meteorites, none were found. Here is a citation of a witness[32]:

> "Early in the morning, when everyone was asleep in the tent, it blew up in the air along with its occupants. Some lost consciousness. When they regained consciousness, they heard a great deal of noise and saw the forest burning around them, much of it devastated."

Figure 20. This map shows the location of three significant events, the most intense experience by planet Earth in recent times: Chelyabinsk, Tunguska, and Kamchatka. (Credit: Russia-CIA WFB Map, Bobby D. Bryant, Wikipedia license, CC BY-SA3.0).

Figure 21. A rendering of the event over Tunguska. (Credit: public domain).

> "The ground shook, and incredibly prolonged roaring was heard. Everything round about was shrouded in smoke and fog from burning, falling trees. Eventually, the noise died away, and the wind dropped, but the forest went on burning. Many reindeer rushed away and were lost."

> "One old man 30 km away was reportedly blown about 12-15m into a tree, causing a compound fracture of his arm, and he soon died. Hundreds of the herders' reindeer, in the general area around ground zero, were killed. Many campsites and storage huts scattered in the area were destroyed."

The researcher Innokenty Suslov[40] cites the accounts of two Evenk brothers, Chuchancha and Chekaren, from the Shanyagir clan, whose tent was located only 20 kilometers from the epicenter.

> "The trees are falling, and their needles are burning, the branches are burning, my herd of reindeer is burning. All around, there is smoke, my eyes hurt. It is so hot, and you could burn. This morning was sunny, there were no clouds, the sun shone brightly, as always, and then, suddenly, a second sun appeared!"

The blast destroyed the brothers' tent, and although they suffered burns they both survived. Thousands of works have been published on the Tunguska event. Four of them pointed out evidence that the Tunguska object belonged to the Taurid Complex[34,35,36,37]. Since this is the most extreme event in recorded history, it implies that the Complex is not as tame and innocent as it was thought to be, but a dangerous congregation of comets and asteroids that could once again strike Earth in the form of a Tunguska II.

Figure 22. According to some estimates, eighty million trees were felled in the event over an area of about 2.500 km². (Public Domain)

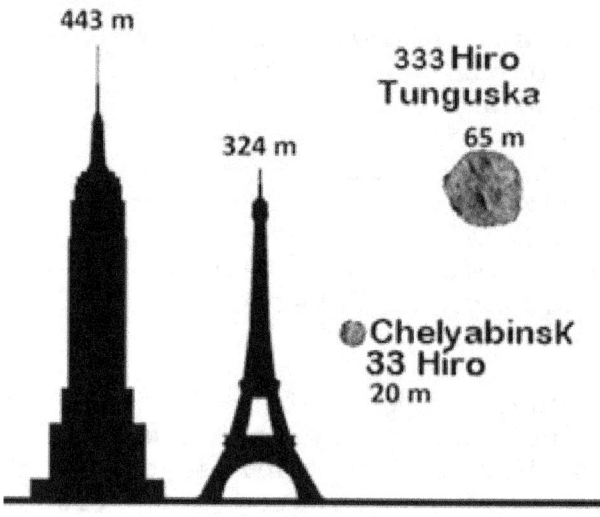

Figure 23. In this comparison, the size of two cosmic bodies with the Empire State Building of New York and the Eiffel Tower of Paris shows that even the smaller object carries a lot of punch. **We define a new unit of Destructive Power, 1 Hiro = atomic bomb of Hiroshima (see Exhibit 33)**. (Credit: Wikipedia Commons, original image modified by IF).

It also tells us that objects of ~100 meters in diameter do exist in the Complex. And as the saying goes: *"If there is one, there could be many."*

Tunguska is an essential object of study because it raised our awareness of the dangers of outer space (Figure 26). The event caused casualties and surface destruction, albeit in small intensity, thanks to the depopulated area in which it fell (Figures 20, 24, 25). But we may not be so fortunate the next time it happens. The bolide moved from SE to NW. If it had fallen one hour and 27 minutes later, it would have affected Moscow, where the number of casualties could have been in the millions.

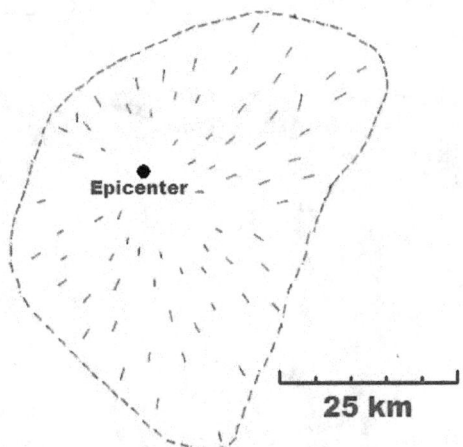

Tunguska Fallen Trees

Figure 24.
The area of fallen trees around the Epicenter of the Tunguska event. Although Kulik expected to find a crater at the center of the distribution, the epicenter, to his surprise there was none (Figure 25).
(IF)

Figure 25. When Kulik visited the epicenter area in 1927, he was surprised to find a forest of "telegraph poles," dead trees still standing. Still, their twigs and branches were blown away. This pattern would take place if the explosion were above the area. Then the blast wave would hit the ground perpendicularly, not felling the trees but stripping them of foliage. (Credit: I. M. Suslow, Moscow, 1928).

Figure 26. Nicolay Fedorov was a well-known Russian artist. This beautiful painting captured the moment of the entrance of the Tunguska objects in the Siberian taiga. Notice the double shadows. The darker one is from the bolide. (Credit: Courtesy Dr. Natalia Fedorova).

2- The Chelyabinsk Event

Early in the morning of February 15th, 2013, several explosions were heard near the Russian city of Chelyabinsk, whose location can be seen in Figure 20. An asteroid of about 19 meters in diameter entered the Earth's atmosphere, exploding at about 29 km in altitude, nearly three times the altitude of commercial airplanes. It had a destructive power of about 33 Hiro (the DP of Hiroshima's bomb) (Exhibit 33). The orbit extends beyond the orbit of Mars but does not reach to planet Jupiter (Figure 29).

Government sensors timed the peak brightness at 09:32 Local Time when people were already in their workplaces. At the blast wave sound, many ran out of the buildings to see what happened.

The explosion can be seen in Figure 27 and an example of the damage in Figure 28. According to government reports, 1613 people needed medical attention due to cuts and wounds from falling glass and trauma from the shock wave. One hundred twelve people had to be hospitalized, two in serious condition. More than 7300 buildings reported glass damage. In the city of Chelyabinsk alone, 3613 apartment buildings had shattered windows and broken glass. Structural damage included the collapse of a zinc factory roof. One hundred eighteen people reported temporary deafness. On a web-based inquiry, 10% of the people reported brain concussion.

This video where the sky trajectory, the explosion, and the arrival of the blast wave have all been registered, is recommended:

https://www.youtube.com/watch?v=dpmXyJrs7iU

The event took place at a distance of 47 km from the city. Had it taken place right above, it is possible to estimate that the damage would have been about 2 1/2 times larger, 18.250 buildings damaged, and 4000 people wounded.

Figure 27. When the asteroid entered the atmosphere, it exploded with a brightness equivalent to 30 Suns, dangerous to stare at. (Credit: Frame extracted from a Youtube Video, Youtube.com/Tuvix72).

Figure 28. The blast wave damaged more than 1700 buildings, like this one. Notice the team of workers repairing the electricity lines. (Credit: Frame extracted from a Youtube Video, Youtube.com/Tuvix72).

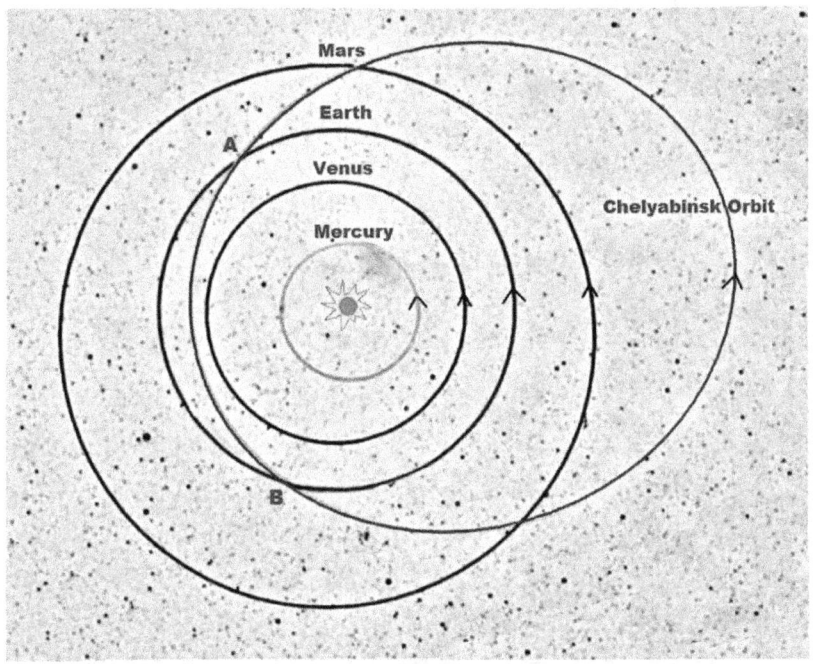

Figure 29. The Chelyabinsk asteroid has an orbit that extends beyond the orbit of Mars. It did intercept the Earth's trajectory at two points, **A** approaching from the night side and **B** approaching from the dayside. Since the event took place during the day, **B** was the point of impact. The image is reproduced as a negative because negatives show much more detail than positives. (Credit: Wikipedia Commons, public domain).

The damage of these two events, Tunguska and Chelyabinsk, was relatively mild due to the unpopulated areas where they fell. But we may not be so fortunate the next time, especially if they fall in densely populated areas like the East Coast of North America, the center of Europe, or China.

The worrying factor is that these two significant events were separated by only 105 years, suggesting that the probability of repeating these events might be of the order of ~100 years, give or take. It is virtually sure that we are going to see a significant event during this century.

Figure 30. The location of Fatima on a map of Portugal. (Credit: Public Domain).

Figure 31. Lucia Santos (10 years old), Francisco (9) and Jacinta Marto (7). Francisco died of the influenza epidemic that followed the First World War in April 1919, one day after taking his first communion. He was only ten years old. Jacinta died on February 20, 1920, from an attack of pleurisy when she was only nine years old. Lucia died in 2005, 97 years of age. The Lady had predicted the early departure of the two younger children. (Credit: Public Domain).

Review

CHAPTER 4
- Image Interpretation

In this Chapter we will analyze some Fatima event images (Figure 30, 33-39) to see if we can learn something from them. Let us remember the Chinese saying: *"An image is worth more than a thousand words."*

1- Introduction

Photography was already well developed in 1917, so it is no surprise that the gathering at Cova had many pictures taken. Some of these images may contain exciting information apparent to a trained eye.

Figure 32. This image was taken while the rain was falling, previous to the event. Notice the opacity of the rain in the distance. (Credit: Judah Bento Ruah, 1917).

Figure 33. This is a remarkable image. I counted 21 people looking up at the same time with no protection whatsoever. Let me ask you this question: For how long can you gaze at the Sun? Not even a second or you may get blind. Thus, so many people looking up implies that the Sun was dimmed, lacking its full luster. Incredibly, of the 70,000 people attending not even one reported eye damage or blindness. This is photographic evidence of the dimming. The second witness from the left is a very tall man. (Credit: Judah Bento Ruah, 1917).

Figure 34. Details of the people looking up. The Sun brightness does not seem to bother them.

Figure 35. This image testifies that the number of people attending the event was in the thousands. (Credit: Judah Bento Ruah, 1917).

Figure 36. Enlargements of faces of the witnesses looking up. The remarkable conclusion about these images is that people are looking at the Sun without protection and it does not seem to bother them. (Credit: Judah Bento Ruah, 1917).

Figure 37. This image shows that the attendance was abundant. (Credit: Public Domain).

Figure 38. People that attended with the children. (Credit: Public Domain).

Figure 39. In the newspaper *"O Seculo"* ("The Century," in English), Avelino Almeida wrote: *"The Sun could be stared at without the least effort. It did not burn or blind."* I counted an additional 13 people looking up. (Credit: Judah Bento).

Figure 40. Enlargements of the people attending. Notice the man in the right pointing to the sky. The witness Monsignor in the book by John de Marchi[4] reported that: *"Arms were raised to point to something above."*

Figure 41. I took this photograph of the full Moon in daylight from Medellín. To be able to see the Sun without protection, its brightness had to be like the full Moon. The Fatima event was not an eclipse of the Sun, but a *"dimming,"* which is quite different, caused by the dust vaporized from the comet. If a photo had been taken during the miracle, the Sun would have looked comparable to this image, not dark, except that it should have exhibited a "halo" of bright comet matter around. (IF).

Figure 42. Some reports have claimed that this photo was taken at the time of the miracle and supposedly is showing the eclipse. A trained eye can conclude immediately that this image was not taken in Fatima. Why? First, because it shows a dark Sun. A dark Sun is only possible during a **total solar eclipse**, but the Fatima event was not an eclipse but a "dimming." Second, because the Sun is near the horizon, while in Fatima it was at ~40° altitude. Where and when was it taken, then? (Credit: L'Osservatore Romano).

In Figure 41 we show an image of the Moon taken in daylight. This photo is a good representation of what was seen in Fatima.

The picture in Figure 42 was published by the Vatican newspaper *"L'Osservatore Romano, 1951,"* purporting to show that it was taken during the Fatima miracle. It has been reproduced on the internet 103 times. According to new information[5], it may have been taken in 1925.

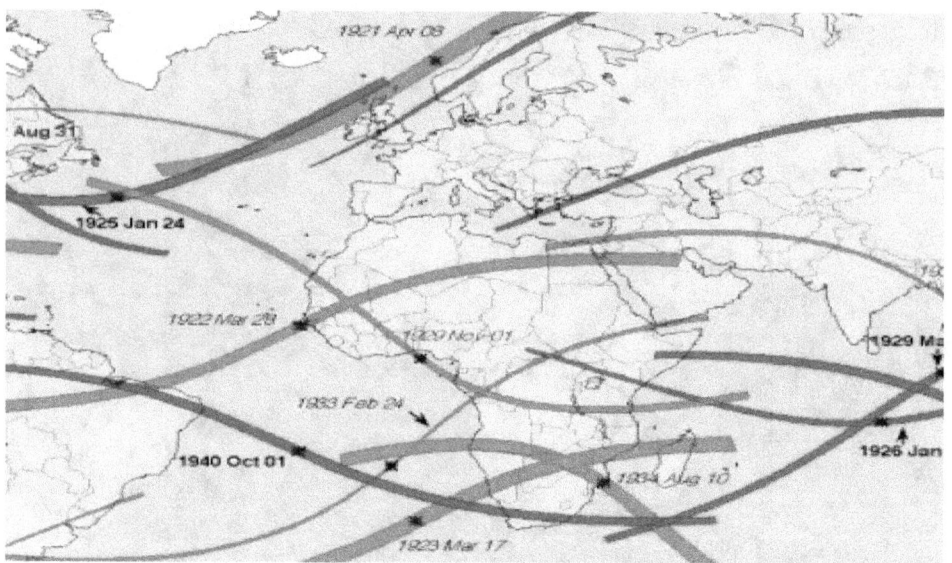

Figure 43. Trajectories of eclipses 1921-1940. The blue lines represent total eclipses, the red lines annular eclipses. There are no total eclipses nearby Portugal. (Credit: Fred Spenak, NASA).

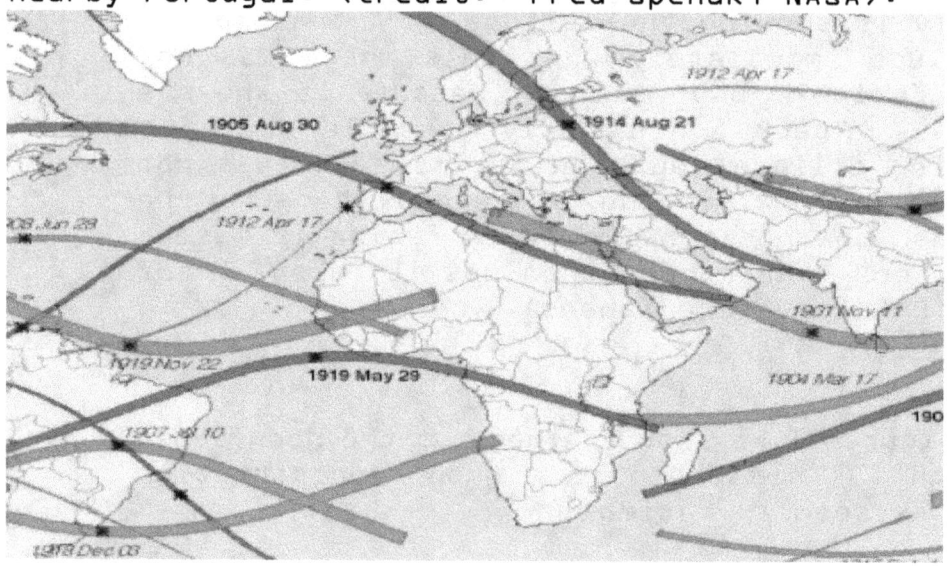

Figure 44. Trajectories of eclipses 1901-1920. There are three total eclipses and a hybrid eclipse nearby Portugal. A hybrid eclipse (total and annular) passes over Portugal in 1912 on April 17, with totality a mere 200 km (125 miles) away. (Credit: Fred Spenak, NASA).

Fred Spenak of NASA[b] has mapped all solar eclipses from -1999 to +3000. We show the paths of all solar eclipses from 1921 to 1940 in Figure 43, while in Figure 44, we show those that have taken place between 1901 and 1920. There were no total solar eclipses in 1925, so the date could not have been 1925.

In Figure 44, covering the period 1901 to 1920, there are three total eclipses not far away from Portugal, in 1905 August 30, in 1914 August 21, and in 1919 May 29, distance 390 km (240 miles), 2980 km (1900 miles), and 4040 km (2530 miles), respectively. There is also an annular-total (hybrid) solar eclipse on 1912, April 17, over Portugal, 200 km (125 miles) away from Fatima[10]. Figure 45 shows the Sun's path in Spain.

With his information, we conclude that the most probable date when the Sun image was taken was during the hybrid solar eclipse of 1912.

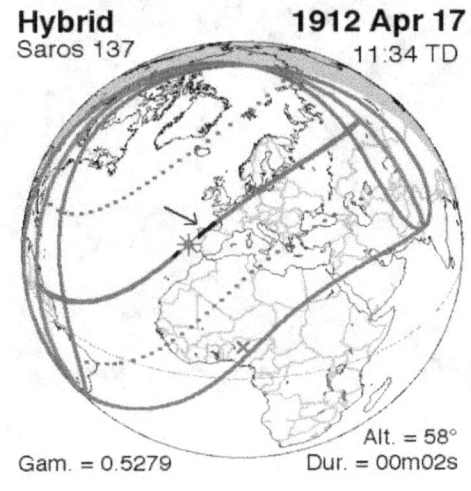

Figure 45. The 1912 April 17 hybrid eclipse had its totality path (arrowed black line) just over Spain and Portugal, and the maximum (star) just nearby over the Atlantic Ocean near the coast. (Fred Spenak and Jean Meeus, NASA).

Figure 46. This comet 2P/Encke image, the parent of the Taurid Complex to which the Fatima Comet belongs, was taken by Michael Mattiazzo, an amateur astronomer from Adelaide, South Australia, with a Celestron 11" Telescope, a Canon 6D camera, and an exposure of 3 minutes. The open cluster of stars above is object 67 in the Messier catalog. (Credit: Michael Mattiazzo).

Research
CHAPTER 5
- Events previous to the Miracle

This Chapter will list some of the events (facts) that took place previous to the Miracle. We have one rule in place: few or no references to religious matters - just detecting facts - to see what we can learn from them. We will use excerpts from the book *"The True Story of Fatima"*[4] by John de Marchi, because he cites Lucia's own words and gives a detailed account of the events from where we want to extract facts. The book is recommended for a more lengthy and comprehensive report.

1- Introduction

The events we will relate, took place in 1917 in Cova da Iria, near the small village of Fatima, where three children declared that they have seen a Lady and had talked to her. The children were Lucia dos Santos and Jacinta and Francisco Marto or ages 10, 9 and 7.

The Roman Catholic Church approved and promoted the Fatima apparitions in 1920, and seven successive Popes had endorsed the apparition and Message.

Previous to the May 13 apparition, it seems that an angel had appeared to the children several times in a prior encounter with the Lady, but these events are independent of the Miracle and will not be considered further.

2- Apparition #1, May 13th

On May 13 (the month of the flowers), the three children had to take the sheep to the field for pasture, and on this day, they decided to go in the direction of Cova da Iria. They reached the place near noon when they heard the church bells summoning the people to the mass, so they knew that it was time for lunch. They opened their bags and ate, and after finishing, they did their Rosary.

De Marchi, in his book[4], writes:

> *"Suddenly they saw a bright shaft of light piercing the air. In their efforts to describe it, they called it a flash of lightning."*

However, De Marchi clarifies in a footnote that *"It was not lightning, but the reflection of a light which approached little by little."* The Lady declared:

> *"I come to ask you to come here for six consecutive months, on the thirteen-day, at this same hour."*

The Lady is setting up the hour of the encounters, and this will also be the hour of the Miracle, past mid-day.

The Lady continued talking on other religious matters and revealed that the three children would go to Heaven when they die. Then She took off to the sky again. Lucia described that *"instead of bodily exhaustion, we felt a certain physical strength."*

Lucia thought it best to keep the apparition a secret, but she doubted that Jacinta could keep the secret to herself. But when she arrived home she ran to her mom crying:

- *"Mother, I saw Our Lady today in the Cova da Iria!."*

Her mother did not believe her, but she kept insisting on what she saw. And Francisco confirmed the account of Jacinta. The father later wrote:

- *"I believed what the children said was true almost at once. Yes, I believed immediately.*

> *Did I think that the children might be lying? Not at all!."*

The news caused such a sensation that it spread all through the village, finally reaching Lucia's family. Her sister was the one to ask:

- *"Lucia,"* she said, *"I have heard people talking saying that you saw Our Lady at Cova da Iria. Is that true?"*

- *"Who told you?"*

- *"Jacinta did."*

- *"I had asked her not to tell anyone."*

- *"Why?"*

- *"I do not know if it is Our Lady. It was a most beautiful Lady."*

It is interesting to notice the attitude of Lucia. *"I do not know if it is Our Lady"* **is a scientific attitude.** Doubt is a component of Science. It also suggests that Lucia was not making it up. If she had wanted to lie, she would have affirmed that it was the Lady for sure.

Lucia's mother worried a great deal about the situation, but despite her attempt to tell Lucia to confess that all was a lie, the child refused to do so.

3- Apparitions #2, June 13th

Before the next apparition on June 13, the news had spread all through the countryside. Some believed it, most did not, to the extent that their neighbors ridiculed the children and their parents.

Meanwhile, Lucia's mother went to talk to the village Pastor, Reverend Manuel Marques Ferreira. He told her to bring the children to him after attending the appointment, because he wanted to interrogate them separately. Then Ti Marto, the father of the two children, talked to the Pastor, who told him: *"If you want to bring the children to me, do it. If not, don't bring them."*

Among the people who believed, Senhora Maria Carreira, later known as Maria da Capelinha (Maria of the little Chapel), would greatly help the children through her kind understanding and helpful assistance. When she learned about the apparitions, she said:

- *"I am going to find out if this is true. If it is, I want to go there. Where is Cova da Iria?."*

Her husband, Senhor Manuel Carreira, told her:

- *"You must be a fool. Do you think you too will see Our Lady?"*

By chance, the great feast of St. Anthony was approaching and was to be held on the same date, June 13th, as the scheduled apparition. All the people were preparing for the feast. Still, Lucia's family did not want to talk about the Lady to see if Lucia would forget her. When Lucia brought it up for conversation, they diverted her mind from it by talking about their plans to attend the celebration. She insisted:

- *"Tomorrow, I am going to Cova da Iria. It is what the Lady wants."*

Instead of the Pastor's recommendation, both families were trying to prevent their going.

- *"Our Lady! What do you mean, silly little girl? No! Tomorrow we go to the feast. Don't you want to get your roll? Besides, there is the band, and rockets, and a special sermon."*

But music and food no longer attracted the child.

- *"But mother, Our Lady does appear at the Cova da Iria."*

- "Our Lady does not appear to you, so it is useless to go there."

- "But She does. Our Lady said that She would appear and She will," Jacinta answered.

- "You do not want to go to the feast?"

- "Saint Anthony is not beautiful."

- "Why?."

- "Because the Lady is more, much more beautiful. I will go to Cova da Iria. If the Lady tells me to go to the feast of Saint Anthony, I will go."

The day arrived, and Lucia was bitter because she felt a lack of understanding from her family and her's mother's cruel opposition. When she went to her mother's room in the morning, she found the room empty, and her mother had already left. In her memoirs, she wrote:

> "I recalled the times that were past, and I asked myself, where was the affection which my family had for me only a short while ago."

Lucia did not know that her mother had made a decision. She had decided that she better follow her. She would hide so she could watch what went on and see if the girl was lying. Also, she wanted to be there to protect the children in case somebody wanted to harm them.

But before that, she decided to go first to Church. On the way there, she met people walking in the other direction. She talked to them.

- "Look here, you are going the wrong way. That is not the way to Fatima."

- "We just came from Fatima. We are going to see the children who saw Our Lady."

- "Where are you from?"

- "We are from Carrascos. Where are the children?"

- "They are in Aljustrel, but they will soon be coming to the feast."

Meanwhile, Lucia had also gone to Church to find some friends that would be willing to come to Cova. She was able to gather fourteen girls that agreed to go along. But Lucia's brother, Antonio, tried to stop them by offering a bribe of a few pennies.

- "I don't care for your pennies," Lucia said. "All I want is to go to Cova da Iria."

So Antonio gave up the attempt. When the fourteen girls arrived at Cova, they were met by a group of women, among them Senhora Capelinha and her crippled seventeen-year-old.

- "I came here with my son John, who had to use a staff to get along. There wasn't a soul around, so we went back to the road. After a while, a woman came from Loureira. She was shocked to see me there because she knew that I was sick and had been confined to my bed."

- "What are you here for?"

- "For the same reason that you came here."

Then she sat down next to her. Then a man came along from Lomba da Egua, and exchanged some words. Next came a few women from Boleiros. I asked if they were running away from the feast. A woman answered:

- "Some people made fun of us, but who cares? We came to see what happens here and find out whether it is they or we who should be made fun of."

Then the children arrived. They all began the Rosary, and as they finished, one girl started the Litany. But Lucia stopped her saying:

- *"Jacinta, Jacinta, here comes Our Lady. I just saw the flash."*

Everyone knelt near the brush. Lucia raised her eyes toward the sky and as if in prayer was heard to say:

- *"You told me to come here today. What do You want of me?"*

Then beguins a conversation between the Lady and Lucia. Among the things the Lady said are these:

- *"I want you to come here on the thirteenth of next month."*

- *"I want you to learn to write and read."*

- *"I will take Jacinta and Francisco soon. You, however, will stay here a little longer."*

The crowd saw Lucia rise quickly to her feet. Stretching our her arm, she cried,

- *"Look, there She goes. There She goes."*

Maria da Capelinha later reported, that there was a sound like a hissing of a distant rocket and a slight cloud, just a few inches away from the foliage, rising slowly toward the East when the Lady left. Others reported hearing something that sounded like a very gentle voice but did not understand what was said.

- *"It was like a gentle humming of a bee."*

When the children came back, some people were mocking them:

- *"What, you are still on Earth! Haven't you gone to Heaven yet?"*

It was a relief when they entered into their homes.

The next day they had to see the priest. The Pastor hoped to settle his doubts.

- *"This could be a trick of the devil,"* he said.

When Lucia left the rectory, she was in a state of turmoil, very uneasy and worried. She began to doubt the manifestations she had experienced. She could not get the doubts out of her mind. Jacinta quieted her:

- *"Lucia, it is not the devil! Not at all! They say that the devil is hideous and that he is under the Earth in Hell. The Lady is so beautiful, and we saw her rise into Heaven."*

Lucia's only place where she could get some peace was in the company of her cousins near the holm oak.

4- Apparitions #3, July 13

As the third apparition was approaching on July 13, Jacinta and Francisco were happily waiting for the event. On the other hand, Lucia was filled with gloom and despair, so much that she decided not to go to Cova da Iria again.

One day the Pastor was talking to Jose Alves, one of the first believers in the apparitions.

- *"It is the invention of the devil,"* he said.

- *"Not at all, Father,"* spoke Alves. *"There is praying at the Cova, and the devil does not like that."*

On the eve of the 13th, Lucia went to Jacinta and Francisco to tell them that she was not going.

- *"We are going! The Lady told us to go there."*

- *"I will speak to Her,"* Jacinta declared, breaking into tears.

- *"Why are you crying?"* Lucia asked.

- *"Because you do not want to go."*

- *"No, I am not going. Look! If the Lady asks for me, tell Her I am not going because I fear She is the devil,"* said Lucia and hurried away.

The next day, when the time to leave for the Cova had arrived, she felt that all her doubts and fears had melted away. With her heart transformed into joy, she ran to her cousin's house to see if they had already gone. She found them kneeling by the side of the bed, crying their eyes out.

- *"Aren't you going?"* she asked.

- *"Without you, we didn't dare to go."*

But realizing that Lucia had changed her mind, they jumped on their feet.

- *"Let's go,"* and they very happily went.

But when they hit the road, they found that it was jammed with crowds of people. They could barely walk because many people stopped them, asking questions and favors to ask the Lady.

Jacinta's mother, seeing all the people going to Cova, went to Lucia's mother and said:

- *"We must go to Cova too. We may never again see our children. What if they kill them?"*

The father of Lucia, Ti Marto, also decided to go. He confessed that when he took to the road, it was crowded. He could not catch sight of the children. When they got to Cova, there were so many people that he could not break through to the place where the children were. Then, somebody pulled his arm and shouted:

- *"Here is their father. Come right here."*

And he was able to sit very close to Jacinta.

They started the Rosary, which was chanted aloud. When they finished the Rosary, Lucia said:

- *"Close the umbrellas, close the umbrellas. Our Lady is coming!"*

Ti Marto declared:

"I saw something like a small greyish cloud hovering over the holm oak. The Sun turned hazy, and a refreshing breeze began to blow. It did not seem that we were then at the height of summer. The silence of the crowd was impressive. Then I began to hear a hum as of a gadfly within an empty jug, but did not hear a word. It seems that it must have been as when people speak on the phone, not that I have ever used a phone. To me, all this was a great proof of the miracle."

Years later, Lucia gave an account of this apparition. Among other things, the Lady said:

- *"I want you to return here on the thirteenth of next month."*

- *"Will You please tell us who you are and perform a miracle so that everyone will believe that you really appear to us?"* Lucia said.

- *"Continue to come here every month. In October, I will say who I am and what I desire, and I will perform a miracle all shall see so that they believe."*

This is the first time the Lady said that she would perform a miracle three months in advance.

Lucia asked for the petitions of the people, but Our Lady answered:

- *"Some I will cure, and others I will not."*

The Lady also asked for the Consecration of Russia. This was going to be one of the issues raised by the apparitions. Since it is a matter of religion, we will not consider it any further.

At this point, something like a rumble of thunder was heard. A little arch that had been set up to hold lanterns trembled as if in an earthquake. Lucia turned around and said:

- *"There She goes. There She goes. She is gone!"*

People surrounded Lucia and asked:

- *"What did the Lady say to make you look so sad?"*

- *"It is a secret,"* she answered.

- *"Is it something good?."*

- *"For some, it is good. For others, it is evil."*

- *"Won't you tell it?"*

- *"No, I cannot tell it."*

The people were closing in on the children. So Jacinta's father was afraid for her safety. He elbowed his way toward the children, picked up Jacinta in his strong arms, and started down the road home. Lucia's father did the same. And a very tall relative took Francisco in his arms. Perhaps this is the tall man seen in Figure 33

It was estimated that about 5,000 people were present.[9]

5- Apparitions #4, August 19th

With the news of the apparitions spreading like wildfire, the number of people attending Cova da Iria increased by the day and by the month. Some were devout, others were curious, and others did not believe and were mocking the children. *"Did Our Lady have sheep? Did She eat potatoes?"* Many wanted to know the secret and offered money and presents. But Lucia always refused to talk. Ti Marto spoke about this:

> *"Some well-dressed gentlemen came only to laugh and make fun of us, who did not even know how to read. Very often, we were the ones who laughed last. Poor things! They had no faith. The children seemed to sense this type of person and would vanish in the wink of an eye."*

The Marto family was more understanding of Jacinta and Francisco than Lucia's family. They ridiculed her frequently, even more than outsiders. Newspapers were not kind to the children either. Some invented stories and complained that *"the priest was setting up a factory of miracles in Fatima."* The children were variously accused of being epileptics, victims of fraud, and greedy. These accounts gave ammunition to the Church's enemies, who calculated that they could destroy the Church if they destroyed the children.

The village of Fatima belongs to the county of Ourem, and at the time of these events, the Chief Magistrate was Arturo De Oliveira Santos (Figure 47), a man of great political power who in his youth, at the age of 20, had abandoned the Church and joined the Masonic Lodge of Leiria. He was also the editor of a local newspaper, and this increased his power significantly. He also founded the lodge at Ourem.

When he heard about the apparitions and the secret, he decided to undermine the people's faith and the Church's power by exposing the children as liars. He had decided to put them in jail. Only one man had the guts to stand up to him in the interest of justice and truth. That man was Jacinta's father, Ti Marto.

Ti reported that the Magistrate had summoned him to appear at the County House with Lucia on August 11th. Lucia asked:

- *"Aren't Jacinta and Francisco going?."*

- "Why should such little children go there? I will answer for them."

Lucia and her father did not wait for Ti and went ahead of him. Finally, they came before the Magistrate.

- "Where is the little boy?," he asked.

- "What boy?" Ti knew that the Magistrate did not know how many children were involved.

- "It is six miles from here to the village, and the children cannot walk the distance."

Then he began to question Lucia, asking her about the secret. But Lucia had decided not to say a word about it.

- "Do the people of Fatima believe these things?"

- "Not at all. All that is just women's talk. I am here at your orders, and I agree with the children."

- "You believe it is true?"

- "Yes, sir. I believe what they say." The Magistrate laughed and dismissed Lucia. He had other plans.

On Monday morning August 13, Ti Marto recalls:

- "I had just begun hoeing my land when I was called home. As I entered the house, I saw a group of strangers standing there. I found my wife in the kitchen, looking worried. She did not say a word but motioned me to go to the front room. I went to the room, and who was there was the Magistrate."

Figure 47. Arturo De Oliveira Santos, the Mayor of Ourem, who had the children kidnapped and imprisoned in August 1917, because they did not want to reveal the secret that the Lady had told Lucia. (Credit: Franciscan Retreats and Spirituality Center, MN, USA).

- "So you are here," Ti said.

- "Yes, of course. I want to see the miracle too."

- "You all go to Fatima and stop at the rectory because I want to ask the children a few questions."

When they got to the rectory, the Pastor was waiting in the office.

- "Who taught you to say the things that you are saying?"

- "The Lady that I saw at the Cova da Iria."

- "Anyone that goes around spreading such wicked lies as the lies you tell will be judged and will go to Hell if they are not true."

- "If one who lies goes to Hell, then I will not go to Hell for I don't lie and tell only what I have seen and what the Lady has said to me. And as for the crowd that goes there, they go only because they want to. We don't call anyone."

- "Is it true that the Lady has confided a secret to you?"

- "Yes. But I can't tell it. But if your reverence wants to know it, I shall ask the Lady, and if She gives permission, I will tell you."

Ti Marto recalls:

- "The children started down the stairs. Meanwhile, the carriage was brought right up to the last step without my noticing. It was just perfect for him, for in a moment, he decoyed the children into it. Francisco sat in the front and the two girls in the back. The horses started trotting in the direction of Cova da Iria."

- "This is not the way to Cova da Iria," Lucia remarked.

In reality, he was stealing the children and was taking them to his own house in Ourem. Having arrived there, he put them in a room and locked the door.

- "You won't leave this room until you tell me the secret," he warned the children. But none of them said a word.

Fortunately for the children, the wife of the Magistrate treated them kindly and with respect. She took them from the room, offered them lunch, and let them play with her children. She also gave them some picture books to look at.

In the meantime, at Cova, the crowd was even more extensive than in July. But the children weren't there, so the people prayed and sang religious hymns around the holm oak. When it became known that the children had been kidnapped, the crowd gasped.

Some witnesses reported that right after a thunder came a flash, and immediately after, they noticed a little cloud, very white, beautiful, and bright. It stayed over the holm oak tree for a few minutes and then left. Some people began to think that the Lady had come, but not finding the children, She returned to Heaven.

People were upset and considered the Pastor and the Magistrate guilty.

- *"The rumor that I was an accomplice to the sudden kidnapping of the children, I repel as an unjust and insidious calumny"*, the Pastor stated.

The next day the children went through an ordeal of relentless questioning. The first to quiz them was an old lady. Later the Magistrate used bribes and gold coins and promised to scold them with punishment, but the children would not give in. The questioning started in the morning and continued into the afternoon. Finally, the Magistrate told them that he would put them into a tank of boiling oil. Jacinta began to cry.

- *"Why do you cry, Jacinta?"*, asked the Magistrate.

- *"Because we are going to die without ever seeing our parents. None of them have come to see us, neither yours nor mine. They don't care for us anymore. I want to see my mother, at least."*

- *"Don't cry Jacinta"*, Francisco intervened.

At that moment they were in jail, and the many men imprisoned there were not unmoved by the little children. They began to sing and dance. One very tall man picked up Jacinta and began to dance with her. But Jacinta thought that dancing was not the right preparation for the apparition. So she ordered the man to stop dancing and began to pray the Rosary. The prisoners also got on their knees. Then a guard came for the children to be interrogated once again.

- *"The oil is already boiling. Tell me the secret,"* asked the inquisitor. But Jacinta remained silent.

- *"Take her away and throw her into the tank"*,

was the order of the inquisitor. The guard grabbed the arm of the child and locked her into a room. The next to be called was Francisco.

- *"Spit out the secret. The other one is already burned up in the oil. Now is your turn."*

- *"I can't. I can't tell it to anyone,"* replied Francisco.

- *"You say you can't. That is your business. Take him away."* And he was taken to the same room where Jacinta was waiting, safe and happy.

Next came Lucia. She was convinced that Jacinta and Francisco had been killed. She thought that she was next to be thrown into the cauldron of hot oil. After her interrogation, they also locked her in the room where the other two waited.

The Magistrate continued pressing the children, but he reached a moment when he realized that he could accomplish nothing. Then out of fear that enraged people could come to Ourem, he put them into the carriage and took them to Fatima.

In the meantime, the people at Cova were asking,

- *"Ti Marto, where are the children?"*

- *"How do I know?"*

He had just uttered these words when somebody shouted:

- *"Look, Ti Marto, look. The children are on the rectory balcony!"*

Ti Marto recalls,

> "I can't say how quickly I got there and swept Jacinta in my arms. I couldn't say a word. Tears ran down

> *my face. Francisco, and Lucia both threw their arms around me."*

Then Ti Marto spoke to the people:

- *"Boys, behave yourselves. Some of you are shouting against the Senhor Prior, others against the Administrator, and some against the Regidor. No one is to blame."*

The Pastor was happy to hear these words. The rage of the people subsided. It was also too late to go to Cova and meet the Lady.

The next Sunday, August 19th, the children had the custom of going to mass and then to Cova da Iria and a nearby village called Valinhos. At about 4 o'clock, Lucia became aware of the signs that always preceded the apparitions: a sudden cooling of the air, the Sun's fading, and the typical flash. The Lady was about to come, and Jacinta was not there. So Lucia sent John to look for Jacinta. The boy did not want to go, but Lucia gave him two pennies, and he started running home. Just after both came back, the Lady appeared. She spoke:

- *"I want you to continue to come to Cova da Iria on the thirteenth."*

Lucia then told Our Lady that there were so many people in disbelief that she wanted to ask Her to perform a miracle so that all may see and believe.

- *"Yes, in the last month in October, I shall perform a miracle so that all may believe."*

This is the second time that the Lady promised to perform a miracle, two months in advance, at Lucia's request.

It was estimated that about 20.000 people were present.[9]

6- Apparitions #5, September 13th

September 13th was approaching. Many thousands of people were believers, but there were also people refusing to believe. Lucia wrote:

> *"When it came time to leave for the Cova da Iria, I left with Jacinta and Francisco, but there were so many people that we could hardly move a step. The roads overflowed with people. Everyone wanted to see and speak to us. There was no human respect in that crowd. Ordinary people, even noble ladies, and gentlemen, succeeded in breaking their way through the crowd surrounding us, fell to their knees, asking that we bring their needs to Our Lady. Many others, unable to get near us, shouted. "Ask Our Lady to cure my lame child." "Ask Her to bring my husband and son from the war.""*

Right there, they could see all the miseries and afflictions of humanity. When they arrived at the holm oak, Lucia started the Rosary as usual. When she finished, the children saw the flash.

- *"What do you want of me,"* said Lucia very humbly.

- *"Let the people continue to say the rosary. In the last month, in October, I shall perform a miracle so that all may believe."*

This is the third time the Lady predicts the Miracle to be performed in October, one month in advance.

- *"Many people say that I am a swindler who should be hanged or burned. Please perform a miracle for all to believe."*

- *"Yes, in October, I will perform a miracle so that all may believe.*

The Lady repeats her promise of a miracle.

- *"Some people gave me these two letters for you and a bottle of cologne," said Lucia.*

- *"None of that is needed in Heaven."*

After the apparition, many people came to the children to ask them questions: *"What did Our Lady look like?," "Was it really Our Lady?," "Tell us everything that happened."*

This is the report of the Monsignor on what happened:

> *"I left, the morning of September the thirteenth 1917, in a slow carriage drawn by an old horse, togo to the place of the apparitions. Father Gois chose a spot overlooking the vast amphitheater of the Cova da Iria. From there, we could easily see, without coming too close, the place where the little shepherds prayed as they waited for the heavenly apparition. At noontime, silence fell on the crowd, and a low whispering of prayers could be heard. Suddenly, cries of joy filled the air, many voices praising the Blessed Lady. Raised arms could be seen pointing to something in the sky, "Look! Don't you see?"*

(Figures 39 and 40 of Chapter 5 show a man raising his hands to the sky). He continued:

> *"I, too, raised my eyes to probe the amplitude of the skies, hoping to see what the other more fortunate eyes were seeing before me. There was not a single cloud*

in the whole blue sky. Yet, to my great astonishment, I saw clearly and distinctly a luminous globe, coming from the East to the West, gliding slowly and majestically through space. My friend also looked up and had the happiness of enjoying the same unexpected but enchanting apparition. Suddenly, the globe with this marvelous light disappeared before our eyes. Some, however, saw nothing."

Some people reported other signs on this day. There was a sudden cooling of the air. The Sun was dimmed, so much that thousands of people could see the stars even though it was mid-day. **Also, there was a rain of iridescent petals that vanished upon reaching the ground."**

It was estimated that about 30.000 people attended the fifth apparition.

7- Apparitions #6, October 13th

Apparition #6 and the last, was going to take place on October 13th. It was expected with great interest because the Lady, three times, had promised to perform a miracle. The number of people attending the Cova de Iria gatherings was increasing every month. Now the people expected in these apparitions reached to the thousands. Nobody wanted to miss it, even those that did not believe it since they wanted to declare it a hoax. The expectation and the anxiety were great. The enemies of the Church were preparing a big celebration after the Miracle was expected to fail.

The children's families were threatened with severe penalties if the promised Miracle did not take place. There was a rumor that there would be a bomb in Cova to scare everyone that went. Some people suggested taking Lucia away and hiding her just in case somebody wanted to harm her. Lucia's mother went to talk to her:

- "People say that we are going to die tomorrow. They will kill us if the Miracle does not happen.

- "If you want to go to confession, mother, I will go with you," Lucia said.

What follows is part of the report that appeared in *"O Século,"* the Lisbon newspaper.

"Along the road, we met the first groups going to the holy place. Many walked more than 10 miles, men and women, most of them barefoot, with the women carrying bags on their heads, while men leaned over their sturdy shafts and carried their umbrellas just in case. Saying their Rosary in a sad rhythm, as if immersed in a dream, they seemed unaware of all that happened around them, disinterested in either the landscape or the other wayfarers. A woman broke out with the first part of the "Hail Mary," the hailing. Her companions took up in chorus the second part, the supplication. There in the open, under the cold light of the stars, they planned to sleep. All night long, the most varied vehicles moved into the town square carrying the faithful and the curious."

"Almost all of them brought besides food a bundle of straw for the animals. The majority of the pilgrims came from many miles around and from the provinces, and gathered about the small holm oak. This was the center of a great circle around which the devout arranged themselves."

The day came, and it was cold and rainy. However, the rain did not stop the people from coming. Many thousands of people came from every region of Portugal. The newspapers sent reporters, and after the Miracle wrote long articles describing the unusual events that took place. For days before the 13th, groups of pilgrims traveled toward Fatima on foot, with bags of food and clothing. People came from Marinha, Monte Real, Cortes, Marrazes, Soubio, Minde Lourical, and Valinhos. People came by foot, donkeys, horses, and carriage.

The rain kept falling as a drizzle, soaking the fields, men, women and children. All night long the rain fell. People came through the roads and hills with a faith that made the rain irrelevant. In the hills, there were thousands upon thousands of souls in prayer. De Marchi gives a piece of important information:

> *"The Miracle promised by Our Lady [was to take place] on the 13th at approximately 1:30 pm according to legal time. But according to the Sun, **this hour would correspond to noon in Fatima, because the Sun at that moment was at its highest point in the sky.**"*

Marchis continues describing the event:

> *"About ten in the morning, skies became overcast. Soon it turned to rain. Sheets of rain, driven by the chilly autumn wind, whipped the pilgrims' faces, drenched the roads, and chilled the people to the bone. While some sought shelter under the trees, others continued their march with impressive endurance." Some estimated the crowd at Cova da Iria to be at least seventy thousand. Dr. Almeida Garrett, a professor at the University of Coimbra, placed the number at 100.000."*

Senhora Maria da Capelina wrote:

"There were so many people even on the twelfth, that we could not hear the din in our hamlet. People expend the whole night in the open since there was no shelter for them."

Ti Marto relates this:

- *"Marto, you had better not go, for you may be mistreated. The children, as they are only children, no one will hurt them. But you are in danger."*

- *"I am going in my good faith. I am not afraid at all. I have no doubts as to the good outcome."*

"The Baroness of Almeirim had brought two dresses for the girls, a blue one for Lucia and a white one for Jacinta. She dressed them herself and placed garlands of artificial flowers on their heads. It made them look like little angels. We left the house under torrents of rain. The road was oozing mud, but it did not keep the women from kneeling before the children. After many struggles and interruptions, we came to the Cova da Iria. The crowds were so thick that it was difficult to pierce through them."

"It was then that a chauffeur took my Jacinta in his arms and, pushing along, opened the way to the posts continually shouting: "make way for the children who have seen Our Lady!""

Figure 48. *"It was then that a chauffeur took my Jacinta in his arms and, pushing along, opened a way to the posts, continually shouting, "make way for the children who have seen Our Lady!"*

Senhora Maria da Capelina describes the scene:

- *"There was a priest close by who had spent the night near the holm oak, and he was saying his breviary. When the children arrived, he asked them about the time for the apparition."*

- *"At noon,"* answered Lucia.

This is an independent confirmation of the time of the event.

The priest looked at his watch and said:

- *"Look, it is already noon."*

- *"Our Lady never lies. Let us wait,"* responded Lucia.

A few minutes went by, and he looked at his watch again.

- "Noon is gone. Everybody out of here! The whole thing is an illusion!"

Lucia did not want to leave, so the priest began to push the three children away.

- "Whoever wants to may go away. I am not going. I am on my property. Our Lady said She was coming. She always came before, and so She must be coming again."

Just then, she glanced toward the East and said to Jacinta:

- "Jacinta, kneel down. Our Lady is coming. I have seen the flash."

The priest was silent.

- "I never saw him again," said Senhora Capelina.

- "Silence, silence. Our Lady is coming," Lucia cried as she saw the flash.

- "What does your Grace want of me?"

And a conversation began between the Lady and Lucia. After a while:

- "There She goes! There She goes!" shouted Lucia.

It was at that precise moment when the clouds were quickly dispersed, and the sky was clear. **The Sun was pale as the Moon. The Sun had taken on an extraordinary color.**

- "Look at the Sun!" she cried out.

In her memories Lucia wrote:

> - "Here is the reason why I cried out to the people to look at the Sun. I was moved to do so under

> the guidance of an interior impulse."

> - "We could look at the Sun with ease," Ti Marto certified. "It did not bother at all. **It seemed to be continually fading and glowing in one fashion then another.** It threw shafts of light one way and another, painting everything in different colors, the people, the trees, the Earth, even the air. **But the greatest proof of the Miracle was that the Sun did not bother the eyes.**"

Ti Marto used to expend his time in the fields, so he knew very well what the Sun's rays could do when it was hot and how you had to avoid the Sun. So he was amazed that this time the Sun could be looked at with no harm whatsoever.

> - *"At a certain point, the Sun stopped its play of light and started dancing. It stopped once more and started dancing again until it seemed to loosen itself from the skies and fall upon the people.* It was a moment of terrible suspense."

Maria da Capelina described it this way:

> - "*The Sun cast different colors, yellow, blue and white. It trembled constantly. It looked like a revolving ball of fire falling upon the people.*"

As the people perceived the Sun falling upon the Earth, some begged for mercy. Others made acts of contrition. One lady was confessing her sins aloud. Witnesses reported that at last the Sun swerved back to its position and rested in the sky. Everyone sighed with relief.

The newspaper "*O Dia*"⁹ reported on October 17th, 1917:

"At one o'clock solar time, midday by the Sun, the rain stopped. **The sky had a certain greyish tint of pearl**, and a strange clearness filled the gloomy landscape, **every moment getting gloomier.** The Sun seemed to be veiled with **transparent gauze to enable us to look at it without difficulty.** The greyish tint of the mother of pearl began changing into a shining silver disc that was growing slowly until it broke through the clouds. **And the silver Sun, still shrouded in the same greyish lightness of gauze**, was seen to rotate and wander like the circle of the receded clouds! The people cried out with one voice. They fell to their knees upon the muddy ground."

"Then as if it were shining through the stained glass windows of a great cathedral, **the light became a rare blue, spreading its rays upon the gigantic nave...** Slowly the **blue faded away, and now the light seemed to be filtered through yellow stained glass.** Yellow spots were falling now upon the white kerchiefs and the dark, poor skirts of coarse wool. They were spots that repeated themselves indefinitely over the lowly holm oaks, the rocks, and the hills. All the people were weeping and praying bareheaded, weighed down by the greatness of the Miracle expected. These were seconds, moments that seemed hours, so fully lived them."

"The people cried with one voice. Thousands whom faith seemed to transport fell to their knees upon

> *the muddy ground. The people were weeping and praying."*

"*O Século*"[7], another Lisbon newspaper, carried a more detailed account of the extraordinary events.

> *"From the height of the road where the people parked their carriages and where many hundreds stood afraid to brave the muddy soil, we saw the immense multitude turn towards **the Sun at its highest, free of all clouds**. The Sun resembled a plate of dull silver. **It could be stared at without the least effort. It did not burn or blind.** It seemed that an eclipse was taking place. All of a sudden, a tremendous shout burst forth, 'Miracle, miracle! Marvel, marvel!'"*

> *"Before the astonished eyes of the people, whose attitude carried us back to biblical times, and who, white with terror, heads uncovered, gazed at the blue sky, the Sun trembled and made some abrupt unheard-of movements beyond all cosmic laws. **The Sun danced, according to the typical expression of the peasants.**"*

Dr. Almeida Garret[4] of the University of Coimbra describes it in this way:

> *"As I waited," he said, "with cool and serene expectation, looking upon the place of the apparitions, and with a curiosity that was fading because the hour was passing away so slowly without anything to arouse my attention, I heard the rustle of thousands of voices. I saw the people stretched out over the large field turnabout from the point upon which their desires and*

*anxieties had converged so far to the opposite side, and they looked up at the sky. **It was almost two o'clock war-time or about noon, sun-time.**"*

This is the third time the time of the event was mentioned.

"The Sun had broken jubilantly through the thick layer of clouds just a few moments before. It was shining clearly and intensely. I turned to this magnet that was drawing all eyes. It looked to me as a luminous and brilliant disc with a bright, well-defined rim. It did not hurt the eyes. The comparison (which I heard while still at Fatima) with a dull silver disc did not seem right to me. The color was brighter, far more active and richer than dull silver, with the tinted luster of the orient of a pearl."

*"Nor did it resemble the Moon on a clear night. **Everyone saw and felt that it was a body with life. It was not spherical like the Moon**, neither did it have an equal tonality of color. It looked like a small, brightly polished wheel of iridescent mother-of-pearl. It could not be taken for the Sun as though seen through a fog. There was no fog at that time. (The rain and the fog had stopped). The Sun was not opaque, veiled, or diffused. **It gave light and heat and was brightly outlined by a beveled rim.** The sky was banked with light clouds, patched with blue here and there. Sometimes the Sun stood out alone in rifts of clear sky. **The clouds scuttled along from West to East without***

dimming the Sun. They gave the impression of passing behind it, while the white puffs gliding sometimes in front of the Sun seemed to take on the color of rose or a delicate blue."

"It was a wonder that all this time it was possible for us to look at the Sun, a blaze of light and burning heat, without any pain to the eyes or blinding of the retina. This phenomenon must have lasted about ten minutes, except for two interruptions when the Sun darted forth its more refulgent, lightning-like rays that forced us to look away."

"The Sun had an eccentricity of movement. It was not the scintillation of a celestial body at its highest power. **It was rotating upon itself with exceedingly great speed.** *Suddenly, the people broke out with a cry of extreme anguish. The Sun, still rotating,* **had unloosened itself from the skies and came hurtling towards the Earth.** *This huge, fiery millstone threatened to crush us with its weight. It was a dreadful sensation."*

"During this solar occurrence, the air took on successively different colors. **While looking at the Sun, I noticed that everything around me darkened.** *I looked at what was nearby and cast my eyes away towards the horizon.* **Everything had the color of an amethyst: the sky, the air, everything, and everybody.** *A little oak nearby was casting a heavy purple shadow on the ground.*

> *"Fearing impairment of the retina, which was improbable, because then I would not have seen everything in purple, I turned about, closed my eyes, cupping my hands over them, cutting off all light. With my back turned, I opened my eyes and realized that **the landscape and the air retained the purple hue. This did not give the impression of being an eclipse.** While still looking at the Sun, I noticed that the air had cleared, and **I heard a nearby peasant say, 'This lady looks yellow.' As a matter of fact, everything far and near had changed now. People seemed to have jaundice.** I smiled when I saw everybody looking disfigured and ugly. My hand had the same color..."*

Dr. Domingo Pinto Coelho[18] writing for *"The Ordem,"* a newspaper of Oporto, wrote these words:

> *"**Blood-red flames sometimes surrounded the Sun. At other times, it was aureoled with yellow and soft purple.** Again it seemed to have the swiftest rotation and then seemed to detach itself from the heavens, come near the Earth and give forth a tremendous heat."*

Another witness was the Reverend Manuel da Silva, who in a letter to a friend wrote:

> "Immediately [after the rain], the Sun came out with a well-defined rim and seemed to come down to the height of the clouds. **It started to rotate intermittently around itself like a wheel of fireworks** for about eight minutes. **Everything became almost dark,** and the people's features became yellow. All were kneeling in the mud."

Inacio Lourenco was a nine-year-old boy living in the village of Alburitel, ten miles away from Fatima.

> *"I was gazing at the Sun.* **It looked so pale to me; it did not blind. It was like a ball of snow rotating upon itself.** *All of a sudden,* **it seemed to be falling, zigzag, threatening the Earth.** *Seized with fear, I hid myself among the people. Everyone was crying, waiting for the end of the world."*

> *"Nearby, there was a godless man who had spent the morning making fun of the simpletons who had gone to Fatima just to see a girl. I looked at him, and he was numbed, his eyes riveted on the Sun. I saw him tremble from head to foot. Then he raised his hands towards Heaven, as he was kneeling there in the mud, and cried out, 'Our Lady, 'Our Lady."*

> *"Everyone was crying and weeping, asking God to forgive them their sins. After this was over, we ran to the Chapels, some to one, others to the other one in our village. The Chapels were soon filled."*

> *"During the minutes that the Miracle lasted, everything around us* **reflected all the colors of the rainbow. We looked at each other, and one seemed blue, another yellow, another red, and so on.** *This increased the terror of the people.* **After ten minutes, the Sun resumed its place, pale and without splendor.** *When everyone realized the danger was over, there was an outburst of joy. Everyone broke out in a hymn of praise to Our Lady."*

> *"As the Miracle came to its end and the people arose from the muddy ground, another surprise awaited them. A few minutes before, they had been standing in the pouring rain, soaked to the skin.* **Now they noticed that their clothes were perfectly dry.**"

Avelino de Almeida, Editor of the newspaper "*O Seculo*"[7], wrote this:

> *"The rain stopped falling, and the dense mass of clouds broke up, and the king-star - of dull silver color - at the Zenith,* **began dancing in a violent and convulsive way while beautiful and rutilant colors covered its surface.** *Miracle, as the people were chanting, of natural phenomena as the wise would say? I do not know. The only thing I can say is what I saw. The rest is to be deal with Science or with the Church."*

He was right. The Church has given its verdict on the event, and now it is time for science to give its ruling.

8- Conclusions

We reach the following conclusions on this and the previous Chapter:

(1) There is a consensus on some basic facts. Different witnesses give the same report, like the change in colors and the dancing.

(2) There are many scientific facts **(in bold letters)** that can be used to do a scientific analysis (Chapters 6 and 7).

(3) Several reports agree on the time of the event.

(4) Images of the event confirm that the attendance was in the thousands. There is no way to calculate a number from the photographs.

(5) Images of the event confirm the rain previous to the event (some analysis by other authors deny the rain). Nevertheless, the rain is irrelevant and does not change or alter our conclusions.

(6) Images of the event show many people looking up to the Sun with no protection whatsoever.

(7) We will not comment upon the many religious comments associated with the above excerpts. To not prejudge over religious interpretations, we will always talk about "the Lady."

(8) An interesting fact is that there are no pictures of the Sun before, during, or after the event. What a pity. The argument was made that the sky was too bright for the camera. This may be a valid explanation. A photographic camera in 1917 must have been a precious and expensive piece of equipment to have, and the instructions to use it must have warned: *"Do not point to the Sun, or the whole film will be veiled."* The risk of burning the camera must have been in the minds of the photographers. The only picture of an eclipsed Sun published by the Vatican newspaper *"The Observatory Romano"* was incorrectly dated, as we have shown in the previous Chapter.

Exhibit #1. Image of the famous comet 2P/Encke taken by NASA's Spitzer Space Telescope, showing the long trail of pebbly debris (along the diagonal line). This material is distributed all along the orbit of the comet. The arrow shows the orientation of the pole. Twin jets of material can also be seen moving away from the equatorial region of the nucleus, spreading horizontally due to the centrifugal force. The comet orbits the Sun in 3.3 years, being the shortest orbital period of all registered comets. Every October, Earth passes through the wake of the comet, the Taurid meteor shower. **It is believed that the fall on Fatima might have been a fragment of comet 2P/Encke.** (Credit: Spitzer/NASA).

CHAPTER 6
- Comet Fatima: Analysis I

Merriam-Webster Dictionary, Definition of miracle:

1: An extraordinary event manifesting divine intervention in human affairs.

2: An extremely outstanding or unusual event, thing, or accomplishment.

In this Chapter, we will scientifically study the famous event of Fatima, Portugal, that took place on October 13th, 1917. We will conclude that it was a small comet that entered our planet's atmosphere, and we will present the evidence to prove it. These results have not been published elsewhere and are presented here for the first time. The children did not have any way of knowing that the event was going to happen on that date and at that time.

1- Historical events

When I was ten years old, I was attending a school of the Marist Brothers, and on May 13th of every year, they took us all to a nearby chapel where we chanted this song:

> On May 13th
> the Virgin Mary
> descended from Heaven
> to Cova de Iria

At that time, I was a child, and I could not have guessed that I would research and decipher the event many decades later. On May 13th, 1917, three shepherd children returned home with the news that the Virgin Mary had appeared to them. The three children were Lucia dos Santos and her cousins Francisco and Jacinta Marto.

The controversial events at Fatima gained fame due partly to three secrets that the Lady had revealed[20]. In her memoirs published in 1930, Lucia revealed two secrets, and a third one was to be revealed by the Catholic Church in 1960. In this work, we will not be concerned with secrets. You can read about them on Wikipedia if you wish and in many books and videos on the internet. Instead, we will concentrate on the "miracle" that took place, and in the scientific facts that we concluded prove its authenticity, the "exhibits."

The miracle was predicted on July 13th, three months in advance:

> Lucia said that on the occasion of the third visit to Cova da Iria, on July 13th, 1917, she asked the Lady to perform a miracle *"so that everybody will believe that you are appearing to us."* In reply, the Lady promised a miracle: *"In October I will tell you who I am and what I want, and I will perform a miracle for all to see and believe."*

Again on August 13th:

> *"Yes, in the last month in October, I shall perform a miracle so that all may believe."*

And once again, on September 13th:

> - *"Yes, in October, I will perform a miracle so that all may believe."*

The prediction was made three times.

2- Previous hypotheses

Wikipedia[20] has a compendium of hypotheses of the Fatima event. Since we want to compare these hypotheses with ours, we will examine them one by one and see if they can explain the "17 facts" listed later on. In Chapter 2, on *"The Scientific Method,"* we explained that the scientist's work is to declare any hypotheses as **TRUE** or **FALSE**. That is our task at hand. Numbers in [square brackets] denote the original reference in Wikipedia.

(Hypothesis 1) Theologian, professor, and priest <u>Stanley L. Jaki</u>:

> *"by divine intervention, a coordinated interplay of <u>natural meteorological events, an enhancement of air lens with ice crystals, was made to occur at the exact time predicted, and this is the essence of the miracle</u> [40]."*

He assumed as a fact that the Sun didn't move since world observatories did not perceive any solar movement. The vast majority of the Earth's populace didn't notice it either. He adds:

> *"A sudden temperature inversion must have taken place. The cold and warm air masses could conceivably propel that rotating air lens in an elliptical orbit first toward the Earth and then push it up, as if it were a boomerang, back to its original position. Meanwhile, the ice crystals in it acted as so many means of refraction for the Sun's rays. Only one observer, a lawyer, stated three decades later that the path of descent and ascent was elliptical with small circles superimposed on it. Such an observation would make eminent sense to anyone familiar with fluid dynamics or even with the workings of a boomerang"* [40], [41].

(H2) Science writer <u>Benjamin Radford</u> maintained:

"The Sun did not really dance in the sky. We know this because, of course, everyone on Earth is under the same Sun, and if the closest star to us suddenly began doing celestial gymnastics, a few billion other people would surely have reported it."

We can easily refute this argument. A selected group of people can only observe a total solar eclipse under the eclipse path. The rest of the world does not know about it. So *a few billion other people would indeed **not have reported it*** if something local had taken place. Radford wrote:

"*<u>Psychological factors such as the power of suggestion and pareidolia can better explain the reported events</u>.*" According to Radford, *"No one suggests that those who reported seeing the Miracle of the Sun—or any other miracles at Fatima or elsewhere—are lying or hoaxing. Instead, they very likely experienced what they claimed to, though <u>that experience took place mostly in their minds</u>"* [8].

Merriam-Webster Dictionary, Definition of pareidolia:

- The tendency to perceive a specific, often meaningful image in a random or ambiguous visual pattern.

- The human ability to see shapes or make pictures out of randomness. Think of the Rorschach inkblot test.

Since most witnesses do not mention having seen a face, a human body, or the Virgin Mary during the miracle, this comment by Radford is irrelevant. He added:

"*It is not clear, <u>and photography from the time of the event does not show,</u>*

that it had been raining as much or as long as was reported" [44].

This can also be refuted. Exhibit 2 and other images in the previous Chapter show evidence of heavy rain. The "duration" of the rain does not change a bit the whole event's results, so this is also an irrelevant comment. Moreover it is false.

(H3) <u>Leo Madigan</u>, a former psychiatric nurse and local journalist at Fatima in the late 20th century, also dismisses suggestions from critics of mass hypnosis and believes that *"astonishment, fear, exaltation and the spiritual nature of the phenomenon explain any inconsistency of witness' descriptions."* Madigan wrote that what people saw was *"<u>the reflection of the Lady's own light projected on the Sun itself</u>."* [43]

It is worth mentioning that *"the reflection of the Lady's own light projected on the Sun itself"* is not a scientific explanation, so this can be rejected outright.

Exhibit #2. This photograph refutes the idea that there was no rain. Additionally, the rain's duration is an irrelevant parameter of the event, since the "miracle" took place after the rain ceased. On the right-hand side, a mode of transportation. (Credit: Judah Bento Ruah, 1917).

(H4) In *"The Evidence for Visions of the Virgin Mary,"* Kevin McClure wrote that the crowd at Cova da Iria might have been expecting to see signs in the Sun, since similar phenomena had been reported in the weeks leading up to the miracle. On this basis, he believes that the crowd saw what it wanted to see.

McClure also stated that *"he had never seen such a collection of contradictory accounts of a case in any of the research that he had done in the previous ten years"* [10].

I want to refute Mr. McClure right here, although he will be refuted again later on: *"The collection of facts compiled by Dr. Chojnowski are consistent with the collection of scientific research found in the comet scientific literature, consistent with the research I have done in my previous 40 years"[24-28] and consistent with the scientific papers cited."*

(H5) Professor of physics Auguste Meessen suggests that the human eye's optical effects can account for the reported phenomenon. Meessen presented his analysis of apparitions and *"Miracles of the Sun"* at the International Symposium "Science, Religion and Conscience" in 2003 [47][48]. While Meessen felt those who claim to have experienced miracles were *"honestly experiencing what they report,"* he stated Sun miracles could not be taken at face value and that the reported observations were optical effects caused by prolonged staring at the Sun.[7]

Meessen contends that retinal after-images produced after brief periods of Sungazing are a likely cause of the observed dancing effects. Similarly, Meessen concluded that the bleaching of photosensitive retinal cells most likely caused the color changes witnessed.

(H6) Steuart Campbell, writing for the edition of *"Journal of Meteorology"* in 1989, postulated that a cloud of stratospheric dust changed the appearance of the Sun on October 13th, making it easy to look at, and causing it to appear to be yellow, blue, and violet, and to spin. In support of his hypotheses, Mr. Campbell reported that a blue and reddened Sun was registered in China and documented in 1983 [11].

(H7) Paul Simons, in an article entitled *"Weather Secrets of Miracle at Fátima,"* stated that it is possible that

> "*some of the optical effects at Fátima may have been caused by a cloud of dust from the Sahara*" [51].

(H8) *Skeptical* investigator Joe Nickell wrote that the "dancing sun" effects reported at Fatima were

> "*A combination of factors including optical effects and meteorological phenomena*, such as the sun being seen through thin clouds, causing it to appear as a silver disc." Other possibilities include an "*alteration in the density of the passing clouds*, causing the sun's image to alternately brighten and dim and so seem to advance and recede, and dust or moisture droplets in the atmosphere refracting the sunlight and thus imparting a variety of colors."

Nickell also suggests that unusual visual effects could have resulted from temporary retinal distortion caused by staring at the intense light of the Sun [6], or have been caused by a sundog, a relatively common atmospheric optical phenomenon [52][12].

(H9) In another document Auguste Meessen[21], professor emeritus of the School of Physics of the University of Louvain, Belgium, proposes another hypotheses:

> "*The hypothesis of an extra-terrestrial intervention may seem to be unrealistic to those who are not aware of the real dimensions of the*

> *UFO-phenomenon. There is ample evidence, however, that it deserves serious attention. It is conceivable, in principle, that apparitions and miracles of the Sun could be produced by "alien visitors" having technical capabilities far beyond our own. We can even imagine that extra-terrestrial civilizations could be interested in performing large scale "psycho-social experiments" without letting us become aware of their procedures, to test our reactions and beliefs in religious matters."*

Astronomer Carl Sagan once said that *"extraordinary claims require extraordinary evidence."* The previous hypothesis lacks the extraordinary evidence.

3- Dust in comets

Our hypothesis depends heavily on the existence of dust in comets because the dust causes the dimming. The gas has a much lower opacity. Thus we will present evidence of the cometary dust that has been preserved on the surface of Mars. In the following pages, we identified images of impacts from which we can learn a lot about the dust in comets. Since the dust is meteoritic, comets will exhibit many of the asteroid's physical characteristics.

Exhibit 3 shows a clean impact on Mars, implying that this was multiple asteroids with the largest fragment around 4 m in diameter. Notice how clean the impact is, suggesting no dust, thus being a bare stone. In Exhibit 4, however, notice that the impact has a lot of dark debris around.

We know that comets are covered by a very dark dust layer of reflectivity around 4%. This dust is as dark as fine charcoal.

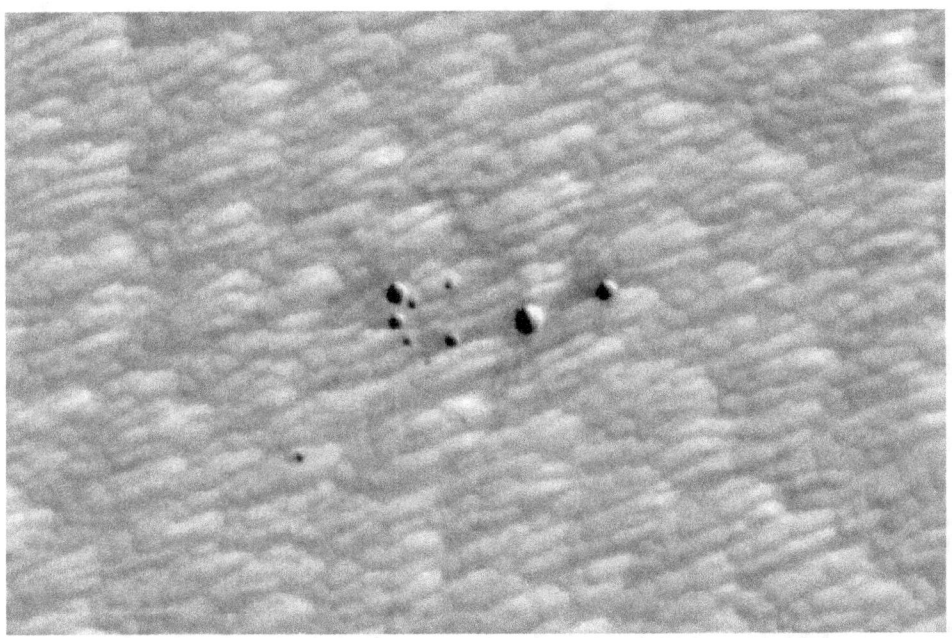

Exhibit #3. The High-Resolution Imaging Science Experiment (HiRISE) camera, aboard the NASA Mars Reconnaissance Orbiter, took this image of Mars showing a cluster of craters created by several pieces of an asteroid. The most significant crater measures about 4 m (13 feet) in diameter. They cover an area of about 30 m (100 feet) in length and are found in a region called Noctis Fossae, located at Latitude -3.213° and Longitude 259.415°. The impacts were so clean that this was a multiple rocky asteroid, not a comet. (Credit: HiRISE/NASA/JPL).

<> <> <> <> <>

In Exhibit 4, the original interpretation of the dark material found around the craters in the literature was that it came from the underlying dark material as a crater ejecta. However, Exhibit 3 was an impact and did not exhibit any ejecta, which implies that ejecta in Exhibit 4 did not come from the inside. It is well known that comets are covered with dark dust material seen in these impacts. Thus Exhibits 4 and 5 show not an asteroid impact but a cometary impact.

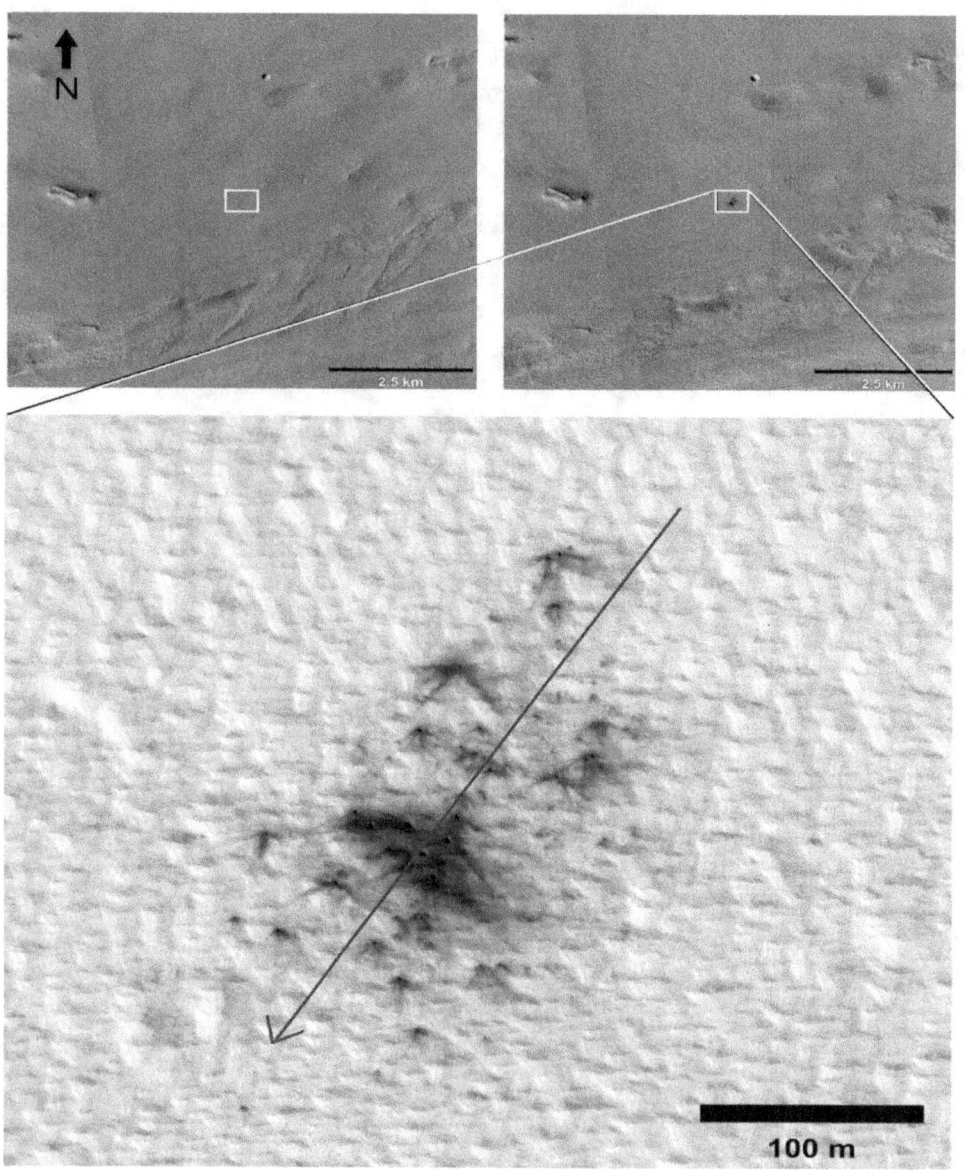

Exhibit #4. This impact on Mars exhibits a lot of dark debris, suggesting that this was a comet impact. Comets are covered with a layer of very dark material reflecting only 4% of the light, and this image is consistent with that knowledge. Notice also that even being a comet, it is composed of quite a few independent fragments (boulders). Typically there is a "seed" of rock around which the ice and dust coalesce. How many fragments can you count? (Credit: NASA/JPL-Caltech).

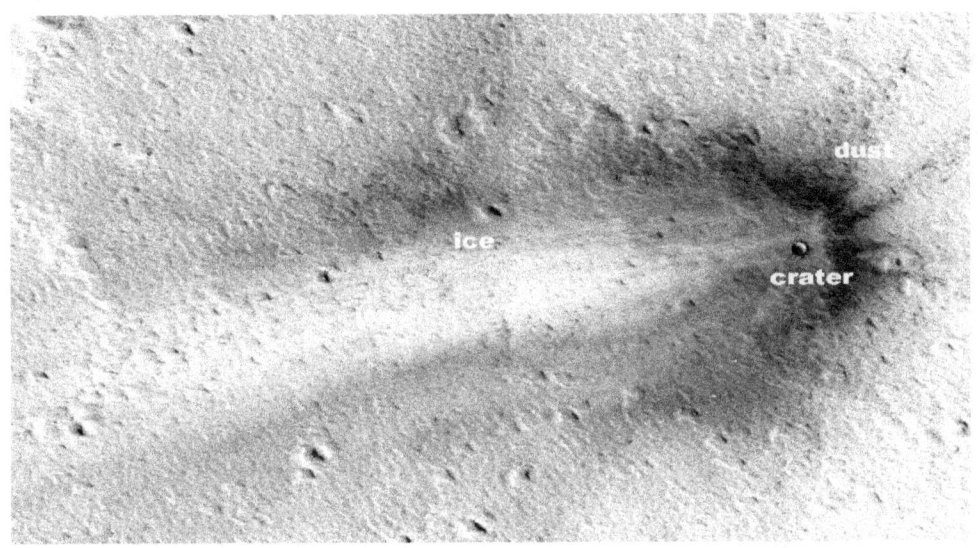

Exhibit #5. This image shows an impact of a comet on Elysium Planitia of Mars taken with HiRISE. Since the atmosphere of the planet is 1/100th of our planet's atmosphere, impacting objects can penetrate to the surface of Mars, with no ablation from the atmosphere. Thus the original objects are preserved. This is an extraordinary image, the **imprint of the black dust** on the surface of Mars that is an abundant component of the comet. The comet has been spread "inside-out," allowing us to see its internal structure: there was a stone at the center of the nucleus (that produced a crater), surrounded by ices (water, carbon dioxide, Carbon Monoxide) and fine carbon dust. (Credit: NASA/JPL, (HiRISE.org/ESP_039148_1980).

<> <> <> <> <>

4. Our Hypothesis

> "Fatima Comet Hypothesis, FCH":
> "The events happening in Fatima on October 13, 1917, were due to a falling upon our planet, of a small comet coming from the Sun's direction, causing a local dimming of the Sun and disintegrating completely in the atmosphere."

5. Analysis

There are many reports and articles on the Fatima miracle containing innumerable facts. We have found that one of the best reports was written by Dr. Chojnowski[12], who gave several talks at the *"Fatima Challenge Conference"*[13] that took place in Rome, Italy, on May 3-7, 2010. His Report is very scientific, consistent, and free from contradictions, and will be the basis of our analysis. Additional information will also be taken from the books by De Marchi[4] and MClLure[9].

Immediately in the introduction, Chojnowski asks the right question: *"What lends credibility to this Miracle of the Sun?"* The Church investigated the events from the beginning, and the Bishop da Silva of Fatima officially approved the apparition as worthy of belief. However, his approval was given using religious arguments, while what we seek here is validation using scientific arguments.

According to the Portuguese historian, Leopoldo Nuñes, who was present at the scene, the witnesses included the Minister of Education, the Masonic Government, illustrious men of letters, arts, and sciences, who came attracted by the prediction. The event was to prove that the Virgin Mary's prediction and her message's authenticity were correct.

One specific witness was Avelino de Almeida, the Editor in Chief of *"O Seculo,"* a masonic daily in Lisbon[7]. The importance of Almeida's testimony is that he was actually there, at the Cova de Iria. The official investigation into the miracle itself began in November 1917, just one month after the event. Monsignor Vidal directed the dioceses of Lisbon to give instructions to start an immediate investigation of the case. Bishop da Silva of Fatima was the one to approve the apparition as worthy of belief. This was not such a difficult decision since the event had innumerable witnesses whose testimonies agreed. So from the very beginning, there was detailed documentation of the event.

On the 60th anniversary (1977), there were still 30 witnesses of the event alive. It has been estimated that 70.000 people witnessed the miracle on the very spot, plus many others far away, 10 to 20 miles from the place.

The Portuguese historian Leopoldo Nuñez who was also present on the scene, said this: *"At the moment of the great miracle, there were present some of the most illustrious men of letters in the arts and the sciences, and almost all were unbelievers, who came out of curiosity. Even the Minister of Education of the Masonic Government was there."*

Dr. Almeida Garret, professor of the Faculty of Sciences at the University of Coimbra, relates that he arrived at Fatima the day before, October 12th, 1917. On the road, he found groups of people on their way to the holy place, 13 miles away. Some men and women were barefoot, with bags. The men carried sticks and umbrellas. At the same time, they were chanting the Rosary in chorus. Since they were reaching the place at night, some slept under the stars.

The day of the event arrived. By 10 o'clock, the sky was thick with clouds, and the rain began to fall in earnest. Although the pilgrims were soaked, nobody left, turned back or complained. They waited for hours near an oak tree where the three children had gathered.

The prediction of the Virgin Mary was specific and even scientific. She made the prediction three times. On July 13th,

> - *"Continue to come here every month. In October, I will say who I am and what I desire, and I will perform a miracle all shall see so that they believe."*

Next, on August 13 and September 13,

> *- "Yes, in the last month in October, I shall perform a miracle so that all may believe."*

Here there is an interesting confusion with the time. Due to the World War I fight with France, the Portuguese government had decided to move the clocks ahead by 90 minutes so that Portugal would be at the same time zone as the army fighting for the French. So, October 13th, 1:30, was actually mid-day, Sun time. The position of the Sun or radiant change very slowly every day. So the time is not critical at all. **Considering all the citations available, the event's best time was midday ±15 minutes.**

About half an hour before mid-day, the three pastors arrived. The little girls adorned with flowers were led to the altar that was erected. Just after arriving, Lucia tells the people to close their umbrellas. The word was transmitted, and despite the rain still falling, everybody closed their umbrellas.

Then they waited.

- Fact #1

Then Lucia glanced toward the East and said to Jacinta[4]:
> *- "Jacinta, kneel down. Our Lady is coming. **I have seen the flash.**"*

A conversation began between the Lady and Lucia. After a while:

> *- "There She goes! There She goes!"*
> shouted Lucia.

It was at that precise moment that the clouds quickly dispersed and the sky was clear. The Sun was as pale as the Moon, and had taken on a unique color. In her memoirs, Lucia wrote:

> *- "Here is the reason why I cried out to the people to look at the*

> *Sun. I was moved to do so under the guidance of an interior impulse."*

- Scientific Interpretation of Fact #1

Fact #1 is that Lucia saw something in the sky and pointed the people to it. What was happening to the Sun that prompted her to point to it?

Dr. Almeida Garret reports[12]:

> *"The sky, which had been overcast all day, suddenly clears up. The rain stops, and it looks like the Sun is about to fill with light the countryside that the wintry morning had made too gloomy."*

At that moment, a diffuse cloud (the comet), coming from the Psi Virginids radiant (Figures 6, 7, 8 and 9), moved in the direction of the Sun, and by the Earth's gravity fell on our planet, actually beginning a "dimming of the Sun" as shown in Figure 6.

In the book written by de Marchis[4], the word used was "flash." In the work by Chojnowski[12], the term used was "lightning." So it is essential to clarify what Lucia saw.

De Marchi, in his book[4], writes concerning the first apparition on May 13th:

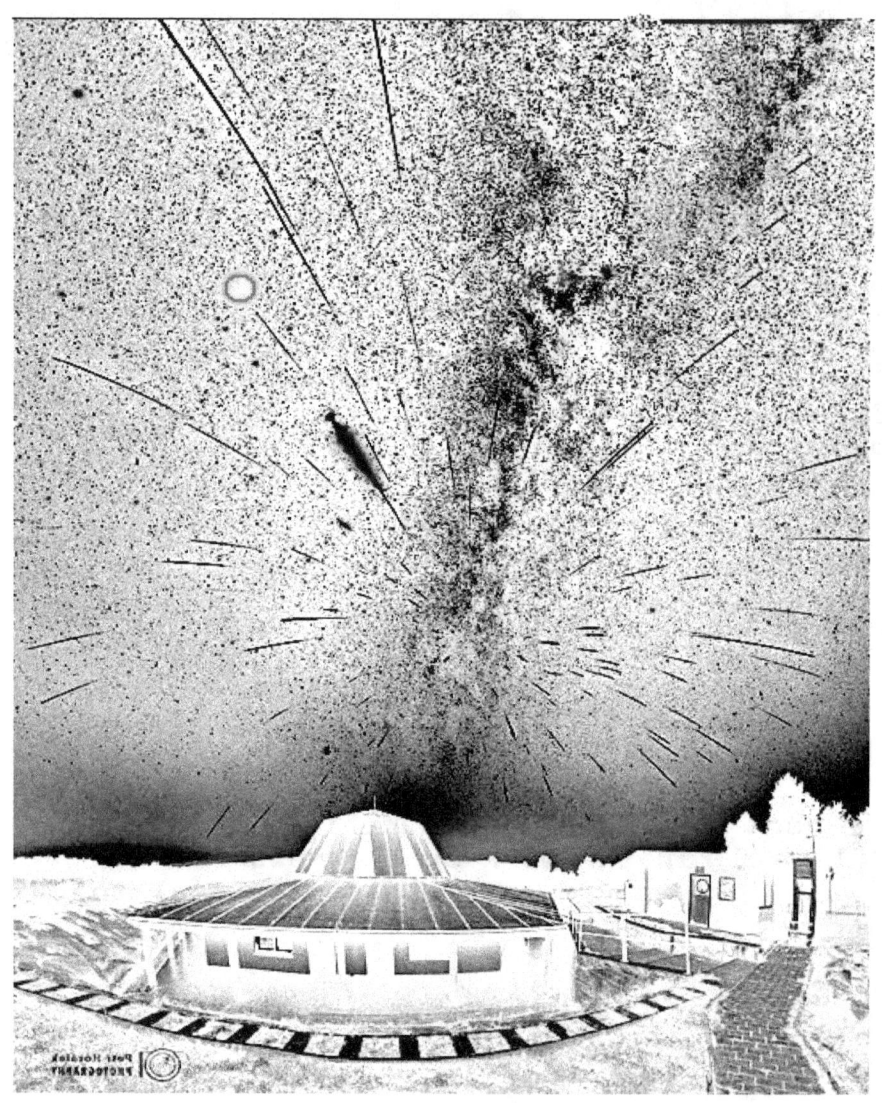

Exhibit #6. The meteor shower of the Perseids in 2013 illustrates the Fatima event. The radiant of the shower marks the origin of all meteors. This work will suggest that the Fatima event was a comet entering Earth's atmosphere (Exhibit 7) from the Sun side (Exhibit 8). The Sun was very near the radiant like in this image and Exhibit 9. A bright meteor moves in the direction of the Sun, and passes over it. We added the Sun and printed the image's negative because negatives show much more detail than positives and because the event took place during the day. (Courtesy Petr Horálek, https://www.astronom.cz/horalek)

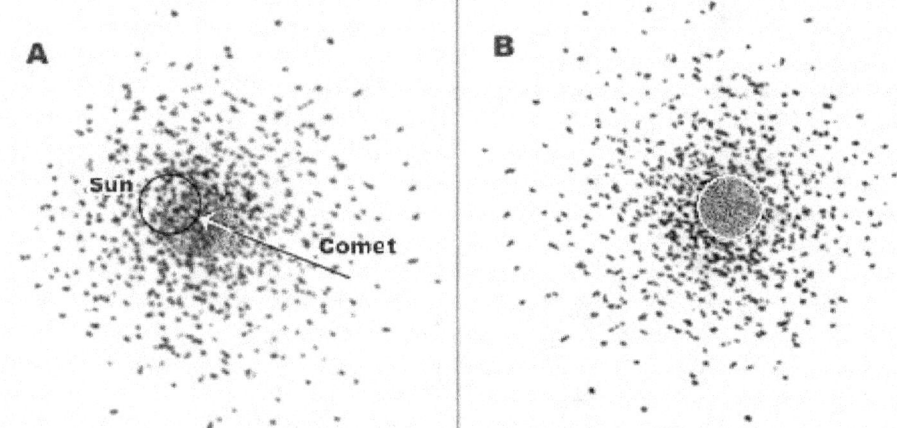

Exhibit #7. This drawing depicts the moment in which the diffuse cloud of dust and gas of the comet starts dimming the Sun (A). It is well known that comets are very dusty, and the dust is expelled from the nucleus because the comet is vaporizing as it enters our warm atmosphere. The cloud is diffuse, but in the dimming (B), the Sun's edge remains sharp. A more complete description appears in the text. (IF).

Exhibit #8. This Figure explains how is it possible to have a meteor shower during the day. The meteors that fall during the day are returning from their encounter with the Sun. (IF).

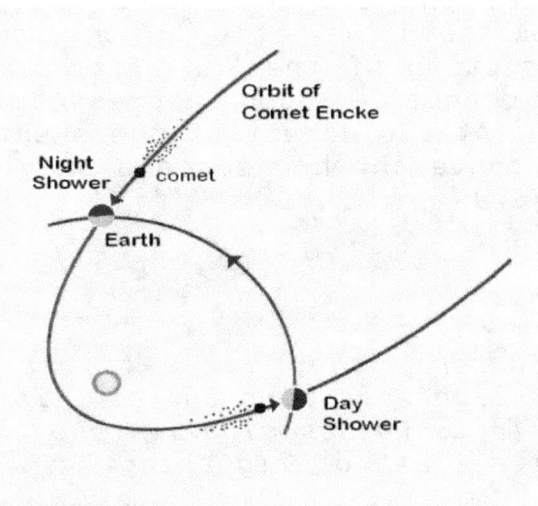

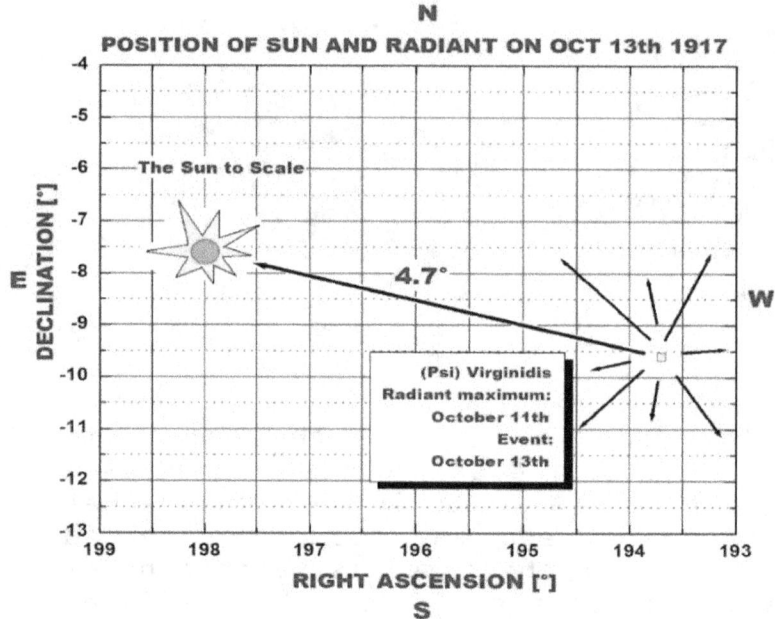

Exhibit #9. In this image, the Sun's position and the radiant on the sky on October 13th, 1917, are plotted with data from the *"International Astronomical Union Meteor Database."* In Chapter 7, we have identified the meteor shower of the Psi Virginids as the origin of the Fatima comet. The thick arrow shows the motion of the comet and clouds in the direction of the Sun, from W to E. Exhibit 6 is by chance a good representation of what happened. At the moment of the event, the Sun was about 40° above the horizon and only 4.7° from the radiant. (IF).

> *"Suddenly they saw a bright shaft of light piercing the air. In their efforts to describe it, they called it a flash of lightning."*

However, De Marchi clarifies in a footnote that *"It was not lightning, but the reflection of a light which approached little by little,"* which is not a scientific explanation.

However, the October 13th apparition is different from the others. The word "flash" is used in previous apparitions in which we cannot identify a miracle. What was the meaning of this word on the day of the miracle?

(1) The comet entrance into the atmosphere is different from a stone's entry into the atmosphere. The stone takes a few seconds to heat up and disintegrate into pieces. On the other hand, the comet, being of density about 1 gm/cm^3, is very fragile. Immediately it enters the atmosphere, it must have a kind of explosion because of the fragility of the body. Maybe Lucy saw this first outburst of light and interpreted that as a "flash." Or,

(2) There is the possibility that the word used by Lucy was actually "lightning," because it is well known that lightning can be seen in the dust clouds emitted by active volcanos (Exhibit 10).

> **Citation from Wikipedia[12]:**
> *"Volcanic lightning is an electrical discharge caused by colliding, fragmenting particles of volcanic ash (and sometimes ice), which generates static electricity within the volcanic plume. Moist convection and ice formation also drive the eruption plume dynamics and can trigger lightning."*

The electric energy generated by the friction of the dust is discharged in lightning. Since comets have a lot of dust, maybe lightning really took place. A fitting description of a comet is "a chunk of very dirty ice." We find that any of the descriptions fit the facts. Then the "dimming of the Sun" started.

- Fact #2

In the book by De Marchi[4] (page 58), Dr. Almeida Garret, professor at the University of Coimbra wrote:

Exhibit #10. Eyjafjallajokull Plume Lightning was observed during the 2010 eruption. Notice the fainter lightning inside the dust cloud, on the right-hand side, above and below. There is the possibility of lightning in a vaporizing comet because it is well known that they are composed of dirty ice, with dust in abundance. (Credit, USGS).

> "While looking at the Sun, I noticed that everything around me darkened."

Ti Marto certified:

> "It [the Sun] seemed to be continually fading and glowing in one fashion then another."

- Scientific Interpretation of Fact #2

As the dust cloud of the vaporizing comet approaches the disk of the Sun, it is increasing in density and thus opacity. So one of the first effects to be noticed by the people is a general darkening of the sky.

- Fact #3

The report by Chojnowski[12] continues:

> "At that moment everyone - and this is an unanimous testimony - **could look at the Sun clearly, directly and without wincing or closing their eyes.** At that moment the multitude turned their eyes toward the Sun."

> "The atmosphere was free from clouds, and the Sun was at its zenith. It resembled a dull silver disc, and it was possible to look at it without the least discomfort."

- Scientific Interpretation of Fact #3

At that moment the small comet had entered the atmosphere from the direction of the Sun, and it was covering its disk with dust producing a solar eclipse of a different type, "a dimming by a dust cloud." The surface temperature of the Sun is about 5700° Centigrade, so the surface is too bright to look at. When the comet enters, at the contact with the atmosphere it begins to heat up and evaporates dust and gases. This evaporation creates an atmosphere (a halo), that subdues the Sun's light. The Sun is seen as if through clouds, smoke or snowflakes. From now on the people are looking at the Sun, but through the dusty comet's atmosphere that is growing extraordinarily in opacity.

- Fact #4

> Dr. Garrett however, had a different interpretation. "I saw the Sun as a disc with a clean cut rim, luminous and shining, but which did not hurt the eyes. I do not agree with the comparison which I have heard made at Fatima - that of a dull silver disc. It was clearer, richer, and brighter in color, having something of the luster of a pearl. I felt it to be a living body. It looked like a

> *glazed wheel made of mother of pearl. It was a remarkable fact that one could fix one's eyes on this brazier of light and heat without any pain in the eyes or blinding of the retina."*

- Scientific Interpretation of Fact #4

Dr. Garrett adds some interesting features. He saw the Sun *"with a clean cut rim which did not hurt the eyes."* Since he is seeing the Sun through a cloud of gas and dust evaporating from the comet, the rim has to look sharp and neat. To not hurt the eyes the brightness of the Sun has to decrease to approximately the brightness of the Moon.

Since the light of the Sun is passing through a cloud of dust and gas, it changed its color. When we see the Sun set, the usual color is red. When we see the Sun through a cloud, the light has a silver or pearl tint, as would be the case if the Sun is seen thorough snowflakes vaporizing from the nucleus of the comet. When we see the Sun through smoke the tint is dark orange.

The next comment is much more significant, *"one could fix one's eyes on this brazier of light and heat without any pain in the eyes."* What is meant by *"Brazier of light and heat?"* When the comet enters the atmosphere it begins to warm up, and this heat vaporizes the ices that compose it. The more it enters the atmosphere the more it heats up, and the nearer it is to the people. Dr. Garrett is beginning to feel this radiating heat.

The final comment is *"I felt it to be a living body."* Since the comet is rotating and sending off mass it is not in a quiet state but actively shedding mass and colors. Thus it could have been described as an "active" and "living body."

- Fact #5

Dr. Almeida Garret[4] of the University of Coimbra:

> "The sky was banked with light clouds, patched with blue here and there. Sometimes the sun stood out alone in rifts of clear sky. The clouds scuttled along from west to east without dimming the sun. They gave the impression of passing behind it, while the white puffs gliding sometimes in front of the sun seemed to take on the color of rose or a delicate blue."

- Scientific Interpretation of Fact #5

Clouds moved from *"West to East without dimming the Sun. While the white puffs gliding sometimes in front of the Sun seemed to take the color of rose or a delicate blue."* The radiant is to the West of the Sun, thus any matter moving from the radiant to the Sun would move from West to East. That is correct.

- Fact #6

Another witness reported:

> "The light was visible 30 miles away, and appeared as a great red flash 16 miles down the road, in Leiria, where the angle of view was such that the climax of the miracle, the "falling of the Sun," was not seen."

- Scientific Interpretation of Fact #6

The flash of light was visible 30 miles (48 km) away, which is a reasonable value because in the case of Chelyabinsk it was seen more than 100 km away, and Leiria is only 17.5 km away from Cova de Iria, or 10.7 miles (not 16 miles).

Meteoroids penetrating the Earth atmosphere have a tendency to disintegrate at altitudes between 70 and 20 km. To do a calculation, let us assume that the explosion took place at 45 km altitude, over Cova da Iria. At Leiria 17.5 km away, the explosion would have happened at an altitude of 31° above the horizon, thus far away from the Sun. This implies that the main part of the phenomenon could not be seen from Leiria. They saw the explosions and the light, but the full occultation of the Sun did not take place from Leiria, or at least not completely. In other words, the event was localized.

- Fact #7

> *"Our Lady wanted to demonstrate Her power in a more convincing way. So this Sun that all the witnesses could observe without irritation "danced." This the common folks described as the 'dance' of the Sun".*

- Scientific Interpretation of Fact #7

The small comet does not have to be perfectly spherical like an orange. It may have the shape of an elongated papaya, and besides, it is well known that comets rotate around an axis. So, when the object enters, the atmosphere may induce some additional rotation. The vaporization of the ice will not be steady, but will produce small explosions or outbursts, and both in turn will produce oscillations in shape and brightness. When the atmospheric pressure reaches a maximum, the whole object may explode once or several times, shedding layers. Dr. Garrett put it similarly,

> *"The Sun trembled, the Sun made sudden incredible movements, outside all cosmic laws. The Sun danced according to typical expressions of the people. It shook and trembled. It seemed like a wheel of fire."*

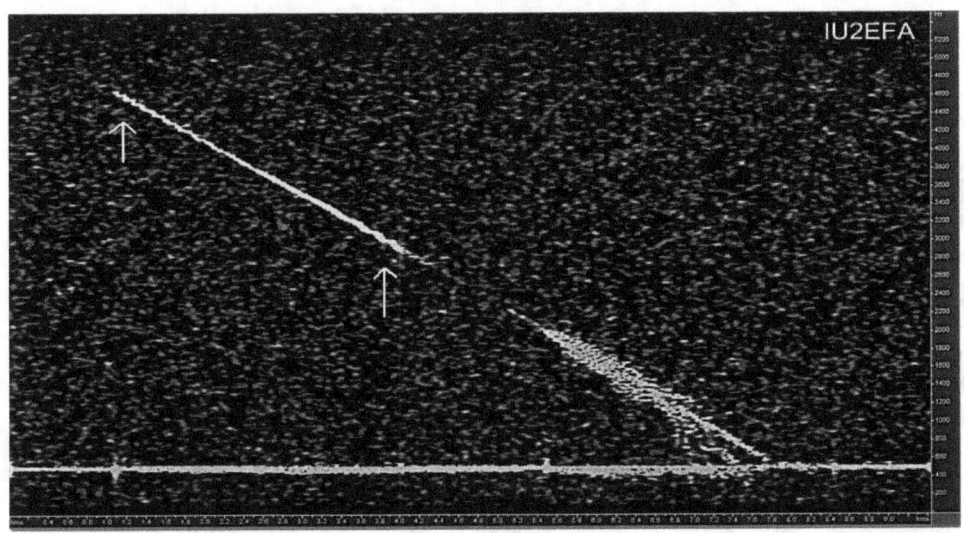

Exhibit #11. The rotating meteor registered on 2020-02-09 10:12 UTC, explains the rotation phenomena observed at Fatima (Credit: IU2EFA, Italy). https://www.youtube.com/watch?v=0E7vVgIFZeI

<> <> <> <> <>

All these facts are correct. Since the object may not be spherical, and since it may rotate, and since it is losing mass, the mass may come apart in different alternative sides giving the impression of "dancing."

"It seemed as a wheel of fire" is a fitting fact. The comet rotates, emits bursts of mass (fire), and looks like a wheel (rotates). This description fits well the entering of a comet in the atmosphere. A rotating meteor entering the atmosphere is seen in Exhibit 11.

- Fact #8

> *"At certain moments the Sun seemed to stop, and then began to move and dance. However, the Sun stops, only to begin its strange dance all over again after a brief interruption whirling upon itself (Exhibit 11), giving the impression of approaching and receding."*

- Scientific Interpretation of Fact #8

"*Whirling upon itself,*" rotating upon itself. "*Giving the impression of approaching and receding.*" As the comet moves down the atmosphere, the front part is heating more and more as it traverses thicker regions. Due to the heat and pressure interactions, it sheds mass that moves toward the observer. The mass is carried over to the sides by the shock wave. This phenomenon happened several times, giving the impression of approaching (sheds mass) and receding (the mass is carried away to the sides).

- Fact #9

"*This dance that was seen by 70.000 witnesses, was actually repeated three times during the course of the 10 minute long miracle*" (Exhibit 12).

- Scientific Interpretation of Fact #9

The dance was repeated 3 times. This is not surprising, and in fact it is also expected. It has been seen many times when meteoroids enter the atmosphere. When the dynamic pressure on the object reaches a certain value, the pressure is enough to break a fragment of the body in an explosion of great magnitude. The body keeps falling at a high speed and a high pressure, and a new fragment falls loose, and then a new explosion. Multiple explosions have been seen many times and there are photometric records showing them. Exhibit 12 shows the disintegration of bolide EN040710 in the atmosphere, with three peaks of large intensity and a few explosions of minor intensity.

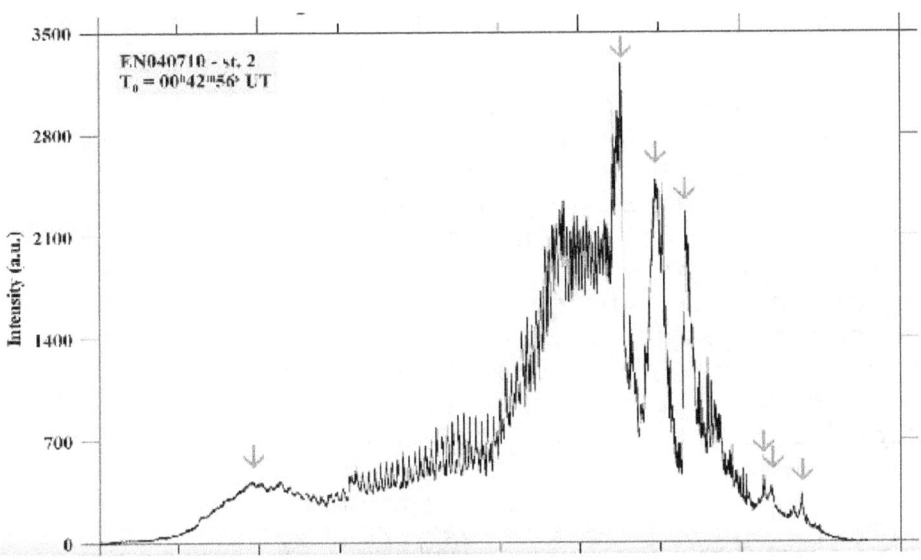

Exhibit #12. Fireball EN04071 light curve, showing three major explosions of light (arrows), and several minor ones like the ones that happened in Fatima. (Credit: Shrbený, L. and Spurný, P.[41]).

Exhibit #13. The explosion of the Chelyabinsk meteor, illuminated the city with an intense, almost monochromatic yellow light, as it happened in Fatima. This is an image extracted from a video uploaded in Youtube. (Courtesy ABC World News).

- Fact #10

Dr. Garrett testified: *"During the solar phenomenon there were changes of colors in the atmosphere. I looked first at the nearest objects and then extended my glance further afield as far as the horizon. I saw everything an amethyst color. Objects around me, the sky and the atmosphere, were of the same color. Soon I heard a peasant who was near me shout out in tones of stupefaction: 'Look, that lady is all yellow.' And in fact everything both near and far had changed to yellow. People looked as if they were suffering from jaundice. My own hand was of the same color."* (Exhibit 13).

- Scientific Interpretation of Fact #10

Comets are composed of many kinds of molecules. The most abundant are water ($H2O$), carbon Monoxide (CO), carbon dioxide ($CO2$), ammonia ($NH3$), methane ($NH4$), cyanogen (CN), sodium (Na), nickel (Ni), and magnesium (Mg). When excited, these molecules exhibit spectral lines in different parts of the spectrum (Exhibits 14 and 15). For example iron and $CO2$ will shine brightly in the visual region, yellow. CN will produce a line in the blue, and ammonia and methane in the red. Sodium will be yellow-orange, nickel green, and magnesium yellow-green and amethyst. Therefore, the sighting of colors can be explained if different layers of the comet are being exposed and excited to brighten depending on the temperature of the object.

The CN Cyanogen molecule is the most intense in all comets. So, when the comet entered the atmosphere, this molecule overcame all others with a deep blue-violet hue (amethyst would be a good comparison), since the band shines at 3883 Angstrom of wave length in the blue-violet region of the spectrum. Actually it is a beautiful almost monochromatic color. Later on, when the temperature of the bolide increased due to the heat generated in the atmosphere, the hue changed to yellow, clearly seen in the original colored images reproduced in black and white in Exhibit 14 and 15.

John Haffert in his book[16] writes:

> "The colors have been characterized as "monochromatic sectors." In other words, the colors were not prismatic, but individual rays of brilliant color."

Fact #4 also referred to the color of the event:

> "The light was visible 30 miles away [48 km], and appeared as a great red flash 16 miles [26 km] down the road, in Leiria."

The key words are *"appeared as a great red flash."* This is what researcher Rubtsov[15] wrote about the Tunguska's light:

> "In the east the brightness of the flying body was much lower than the Sun. Its color was red, and the shape was that of a ball or "artillery shell" with a long tail. Eyewitnesses said simply: a "red fiery broom" or a "red sheaf" was flying, and it was swiftly moving in the western direction, leaving no trace behind. The duration of this phenomenon did not exceed a few minutes."

The color of the light and the duration of the event match perfectly what was seen at Cova da Iria. The observed colors, yellow and red, match perfectly what was seen in Tunguska and Chelyabinsk. These cities did not see the amethyst color because they were not comets, but asteroids.

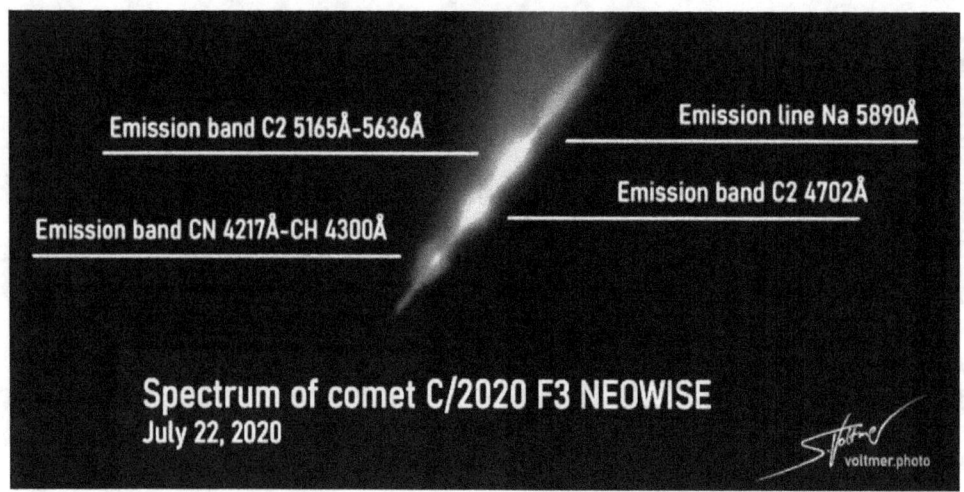

Exhibit 14. This low resolution spectrum of comet C/2020 F3 NEOWISE shows the basic colors that can be seen in a comet, from blue to red in the original color image. For example the blue CN color is produced in a range of less than 20 Angstrom, making it very monochromatic and thus pure and bright. Asteroids do not have the CN band. (Credit: Dr. Sebastian Voltmer, astrofilm.com).

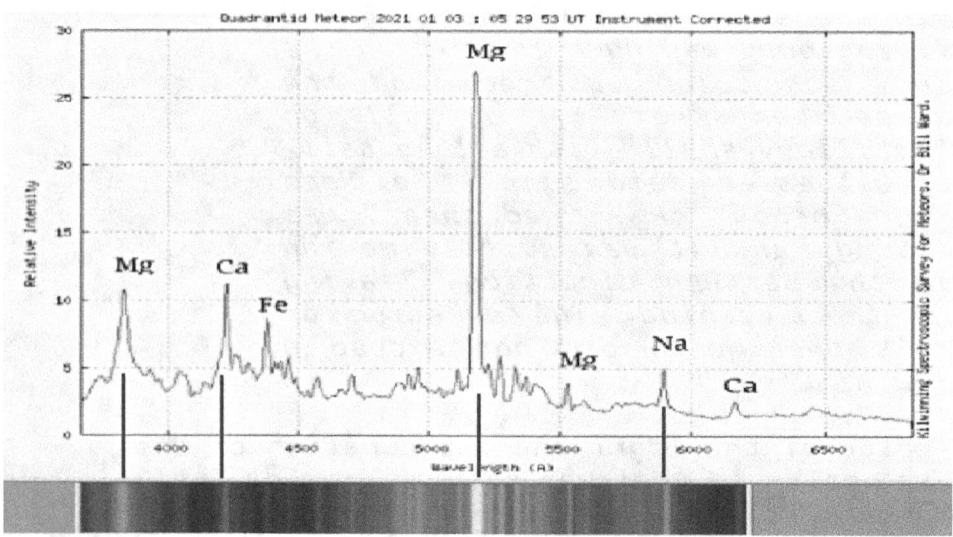

Exhibit 15. This is a spectrum of a meteor of the Quadrantids' shower, secured on the night of January 3rd, 2021. It shows the amethyst color that was seen in the Fatima event. (Credit: Dr. Bill Ward, Kilwinning Spectroscopic Survey for Meteors/International Meteor Organization).

Asteroids do not have a content of the CN Cyanogen molecule which is the most abundant in comets, so they cannot exhibit intense blue color.

- Fact #11

Marco Daniel Duarte[2] found that one of the persons that was present on October 13 was Maria Bettina Basto (1897-1987). In a small notebook she made a sketch of what she saw (Exhibit 16), with a written note on the next page. Duarte registered what was on the yellow pad notebook:

> "The Sun appears among clouds, but with no brightness. It looks like the Moon with a cross in the middle, and then, how marvelous, the atmosphere takes the colors of the rainbow, white, then orange, green, then blue, pink, and lastly golden. Everything is tinted golden. What a beautiful thing. And later on, the Sun completely black, looks like it wants to detach from the sky, and starts rotating and falling."

- Scientific Interpretation of Fact #11

As it enters the atmosphere, the opacity of the dust cloud is increasing all the time. Since the comet will disintegrate completely, the opacity has to reach a maximum (the black Sun), and then it decrease again and clears up completely. The moment of maximum opacity seems to have been registered in the sketch and side note. All the colors mentioned can be seen in Exhibits 14 and 15.

- Fact #12

> "Now the final part of the event, the falling of the Sun. The people present at Cova de Iria, began to believe that the world was ending, so authentic was the experience."
> "Suddenly a scream came from the crowd, a clamor, a cry of anguish, from all the people. The Sun

whirling wildly, seemed to loosen itself from the firmament and advance threateningly upon its huge and fiery weight. The sensations during these moments were terrible."

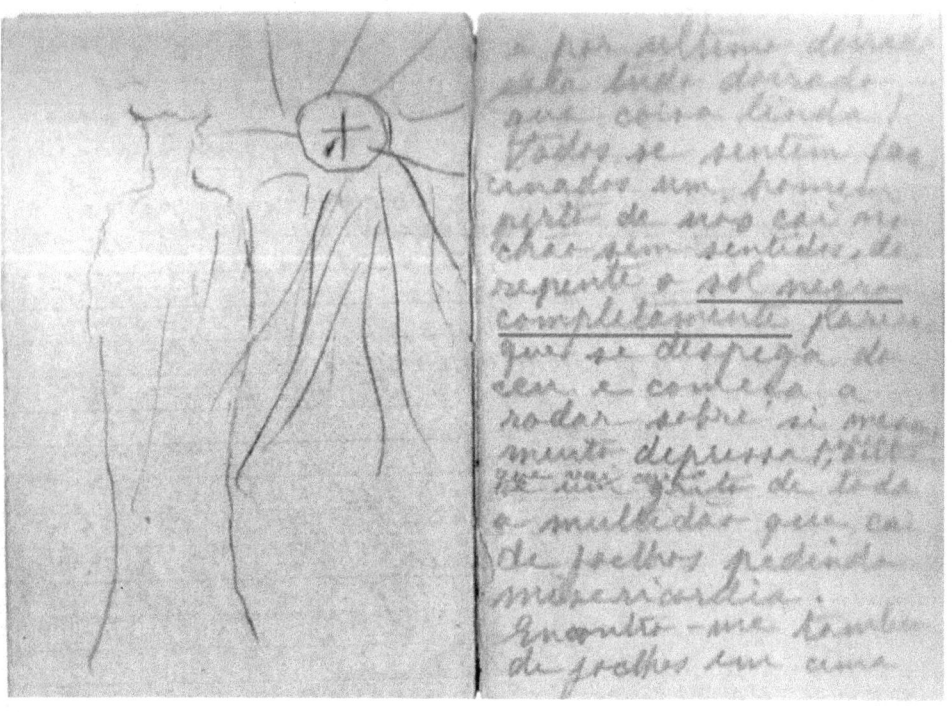

Exhibit 16. This is a copy of the notebook used by Maria Bettina Basto to sketch what she saw on the Sun, written in Portuguese. We ignore the human figure on the left. She mentioned the existence of a cross inside the circle of the Sun, but this must have been an ephemeral design caused by the passing clouds. The black spot on the Sun could have been an attempt to paint the Sun dark to be in agreement with the text on the right. Two scientific facts can be extracted from this evidence. 1- Note the rays coming out of the Sun in all directions, and 2- note that she wrote with her own hand "<u>the Sun completely black</u>." (Credit: Maria Bettina Basto, Marco Daniel Duarte, Carlos Cabecinhas, FATIMAXXI[2]).

Another witness described the scene:

> "Many began to confess their sins, aloud and crying to Heavens for mercy. Parents protected their children. Many were paralyzed with fear. The Sun began to move as if being detached from the sky and was falling upon us. It was as if a wheel of fire was about to fall on the people."

Inacio Lourenco was 10 miles away from Fatima and wrote this:

> "I was gazing at the sun. It looked so pale to me; it did not blind. It was like a ball of snow rotating upon itself. All of a sudden it seemed to be falling, zigzag, threatening the earth. Seized with fear, I hid myself among the people. Everyone was crying, waiting for the end of the world."

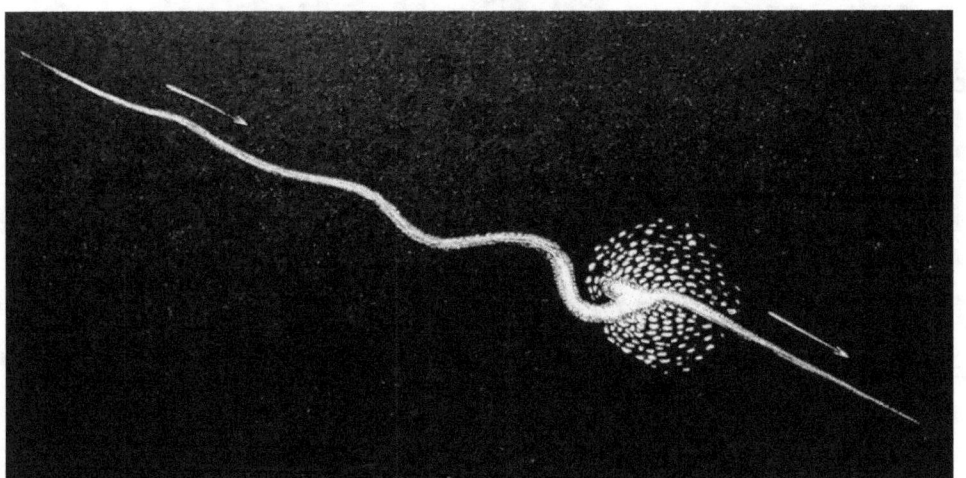

Exhibit 17. This bolide with a zigzag trail was recorded by M. Joseph Silbermann, on November 11, 1869. His drawing represents very well the observation made at Fatima (the dancing), including the explosion. (Credit: Wikipedia Commons, public domain).

- Scientific Interpretation of Fact #12

"The Sun whirling wildly, seemed to loosen itself from the firmament and advance threateningly upon its huge and fiery weight."

The comet was reaching to the end of its trajectory in the atmosphere. Now it may have been at only 10 to 20 km above, just a few seconds from hitting ground. It was actually falling, and that is what the people saw.

"The Sun began to move as if being detached from the sky and was falling upon us. It was as if a wheel of fire was about to fall on the people."

That was precisely what was happening. Indeed the comet was falling after having traversed most of the atmosphere. So it had to be hot, fiery and falling.

In Exhibit 17 we present the observation of a bolide with a wavy trail and explosion, observed by M. Joseph Silbermann, on November 11, 1869. His drawing represents very well the observation made in Fatima, with zigzag included in the trail.

- Fact #13

"Many interpreted this sign as a warning and an indication of Mercy of God, because the Sun returned to the heavens."

- Scientific Interpretation of Fact #13

"Because the Sun returned to the heavens." Having traversed the whole atmosphere, the comet must have exhausted all of its mass, vaporized in the generated heat. Then becoming transparent, the Sun could be seen again, and thus the Sun "returned" to the sky (Exhibits 18 and 19).

- Fact #14

The following quote talks about tangible events[12].

> "Many of the witnesses, after panic was followed by sudden relief, would have thought the whole thing a dream if each one had not had a personal tangible evidence that they had been witnesses of an "explosion of the super-natural.""

> "Many of the witnesses experienced panic and then a sudden relief, since no harm happened."

- Scientific Interpretation of Fact #14

The remarkable thing is that there were no casualties. A comet had disintegrated over 70.000 people, and by chance, all the comet was consumed in the atmosphere and no damaging residue arrived on the ground, saving thousands of people.

Talking of damage, there is another remarkable effect: After 70,000 people looked at the Sun with no protection whatsoever, nobody reported blindness. I searched more than 1600 pages of documents, and I could not find one single report of eye damage. Extraordinary!

- Fact #15

In the book by John M. Haffert *"Meet the Witness"*[16], Dominic Reis gave the following account:

> "As my father described it: There was a good three inches of water where I stood, and mud on the ground. Yes, three inches of water on the ground. I was soaking wet."

> "We put the umbrellas down. . . . Lucia starts to talk to the people, one here, one bunch there, and some

started to kiss the little kid, and the wind started to blow real hard."

"As soon as the Sun went back to the right place, the wind started to blow real hard, but the trees didn't move at all. **The wind was blow, blow and in a few minutes the ground was as dry as this floor here.** *Even our clothes had dried. We were walking here and there, and our clothes . . . . we don't feel at all.* **The clothes were dry and looked as though they had just come from the laundry."**

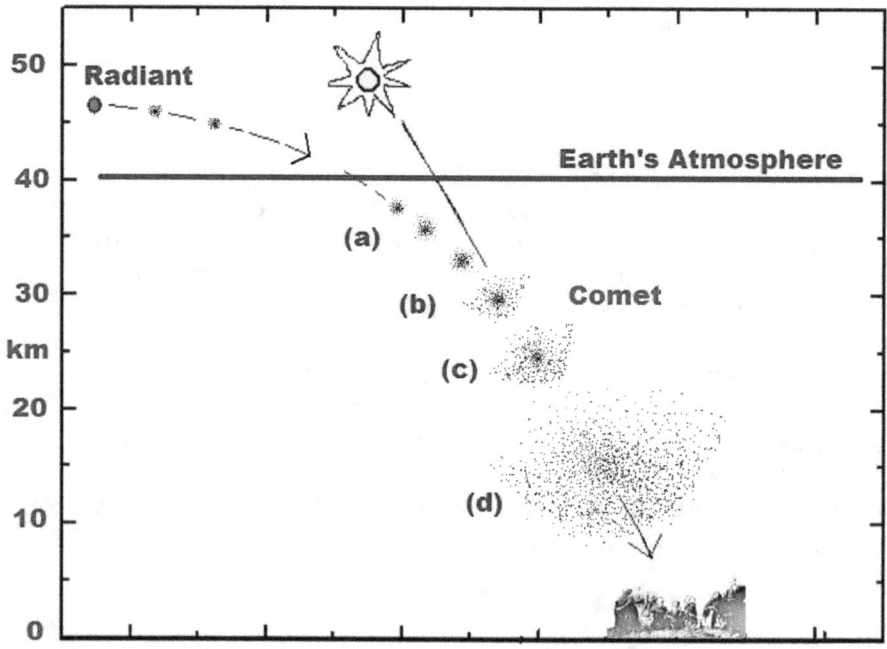

Exhibit 18. Schematic diagram of the entrance of the fragile comet in the atmosphere. After entering (a), it begins to expand due to the vaporization of the nucleus shedding gas and dust, initiating the dimming of the Sun in (b), (c). When it gets to deeper layers, the whole comet is consumed in the atmosphere, becoming transparent again (d), and the Sun returns to the sky. This explains another observed phenomena, the moving of clouds from West to East (Exhibits 6 and 9). (IF).

- Scientific Interpretation of Fact #15
    In this fact we consider only the wind blowing. What makes this paragraph so unique? The phrase "*an explosion of the supernatural*," and the words "*each one*" and "*tangible.*"

    It is well known[17], that whenever a bolide enters the atmosphere, it will explode under the atmospheric pressure, and will send a **sonic wave** or **air blast** in the direction of the impact. It was this wave that killed scores of deer in the Tunguska event and felled 80 million trees, and the wave that broke windows in the event of Chelyabinsk.

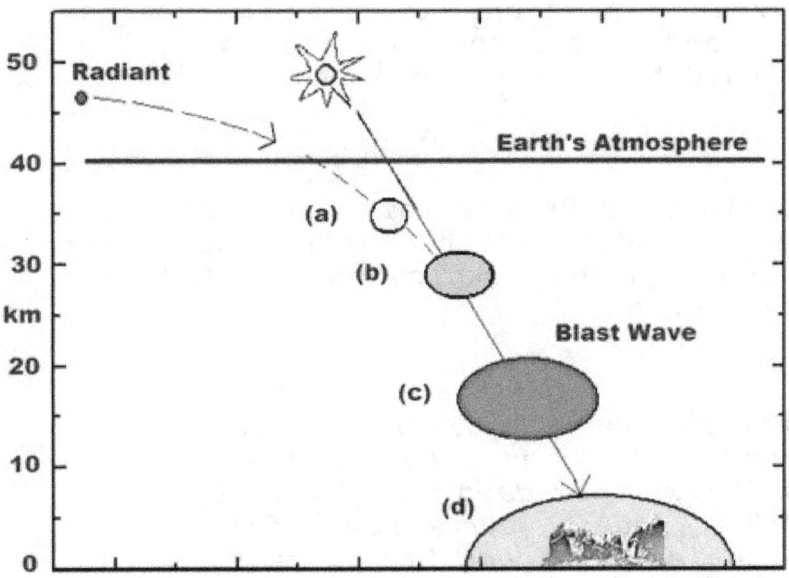

Exhibit 19. Schematic diagram of the descent of the air burst over the people of Fatima. When the comet coming from the radiant enters the upper atmosphere (a), it starts to expand by the atmospheric pressure and the blast wave is generated. The dust descends on the surface causing a dimming of the Sun, (b) and (c), at the same time that it gets hotter and expands. When it reaches the ground (d), it becomes a bubble of hot air that sticks to the surface for some time. That explains the wind that was felt by many people and the drying of the garments. (IF).

Reporting this feature of the event, *"and the wind started to blow real hard,"* and *"the wind was blow, blow and in a few minutes the ground was dry as this floor here. Even our clothes had dried,"* is the best demonstration that it was an atmospheric impact that produced it, and a good proof that the event was real. They could not have made it up because in 1917 there were no atomic bombs, and the concept of a blast wave had not yet been coined. Additionally, notice that to be able to dry the garments the wind had to be hot and dry like in a blast wave.

In the following video available on the internet, several explosions of the Chelyabinsk object can be heard. At second 15 there is a loud explosion followed by 11 minor ones:

https://www.youtube.com/watch?v=dpmXyJrs7iU

Again at minute 3:08 the main blast wave can be heard very loudly, with minor explosions following,## and a strong explosion at 3:27 near the end of the record.

- Fact #16

> *"This enormous multitude was drenched, for it had rained unceasingly since dawn (see Exhibit 2). But now, everybody felt comfortable, and found his garments quite dry, a subject of general wonder. There was no doubt of this fact due to the amount of people that experienced it."*
>
> Dr. Pereira Gens described it thus: *"I still remember the delicious sensation that this warm caress of the Sun gave me. I felt my clothes almost dry now, although they were all wet a few moments ago."*

From Fact #16 we take the quote:

> *"The wind was blow, blow, and in a few minutes the ground was as dry as this floor here. Even our clothes had dried. We were walking here and there, and our clothes . . . . we don't feel at all. The clothes were dry and looked as though they had just come from the laundry."*

- Scientific Interpretation of Fact #16

The total mass of the comet had vaporized in the atmosphere, but the energy of the re-entry remained there in the form of a cloud of hot air, heated along the path of fall, and this cloud was moving toward the gathering of people (Exhibits 18 and 19).

> *"I still remember the delicious sensation that this warm caress of the Sun gave me."*

Delicious it must have been, after the fear they experienced, because it was a mass of warm and dry air that must have come down like a breeze, staying there for quite a while, sufficient to dry their garments and the wet ground.

A similar experience took place at the Chelyabinsk event[41]. Of 1113 respondents to an internet survey who were outside at the time, 25 were sunburned (2.2%), 315 felt hot (28%), and 415 (37%) felt warm, thus confirming the experience. The values depended on the distance of the observer from the explosion.

M. Kosolapov, a witness to the Tunguska event, wrote[15]:
> *"It felt as if someone had burned my ears. A hot wind blew past us."*

The Tunguska event was more powerful than the Fatima event, explaining the *"past us."*

This reminds me of some hot days I spent in Tucson, Arizona. Temperatures of 38°C = 100°F, were common during the summer time. I used to go to the rest room to wet my face and head and then return to work. For how long did this humidity lasted? Not long. Just a few minutes, no more, and I had to do it again to fight the heat. A temperature of 38°C is sufficient to dry your clothes and garments in a matter of minutes.

- Fact #17

Antonio Marujo[14] writes that Maria do Rosario wrote a book with the title *"The Lady of May"* (in Portuguese), where she gave the following report as a witness:

> *"I was there and I saw everything. The Miracle of the Sun made many people cry and experience fear. The Sun began to wander. It looked like a wheel of fire, rotating, rotating, toward the West, coming toward the ground, descending. People became very afraid because they thought that the world was going to end, and the Sun descending, descending. But nothing happened and Lucia said: 'Don't be afraid. This is the miracle of the Sun.' And it was true. The event ended and there was no harm. Then came the rain, very soft and small. White drops began to fall. They looked like flowers of 'gipsofila,' so white they were falling."*

- Scientific Interpretation of Fact #17

At this moment of the event, the Sun has "returned" to its place, which means that the comet must have disintegrated completely or almost completely, becoming transparent again and letting the Sun shine with all its original brightness.

The question then is if the disintegration was full and complete, or if there was a residue that could fall upon the people. Maria do Rosario reported that there was *"white soft rain"* that resembled flowers of *"Gypsophila"* falling on the people (Exhibit 20). The flowers look like cotton balls, small snowflakes, or snow balls.

This fact might be real on two counts. First, snowflakes or snow balls is what we could expect as a residue of the comet. And second, the validity of this Report depends entirely on the timing with respect to the other events. If this Report had come at the beginning of the miracle, we would have to disregard it completely. Instead, the Report comes at the right moment, when we expect the comet to be reaching the ground after the Sun is seen back to its place. In conclusion, the Report looks real and suggests that some residue might have reached the ground.

Exhibit 20. White "Gypsophila A1 Snowball" flowers look very much like cotton balls, snowflakes or snow balls. Apparently some soft rain drops had this appearance when dropping from the sky. (Triangle Nursery, UK).

5. Conclusions of the analysis

(1) To uncover these 17 facts I had to search carefully and systematically through about 1600 pages of documents.

(2) None of the original hypotheses - (H1) to (H9) - is capable of explaining Facts 1-17, so we must conclude that these hypotheses are **FALSE**.

(3) The **"Comet Fatima Hypothesis"** is capable of explaining Facts 1 to 17, so we have to conclude that the hypotheses is **TRUE**.

(4) The children could not have known that the event was going to happen **on that date and at that time**. Someone had to tell them, the Lady.

(5) The event was real, not a hoax. It was not the result of mass hysteria or mass hallucinations. People did not see *"what they wanted."* It was the descent of a small comet that disintegrated in the atmosphere, in a trajectory coming from the Sun, and directed toward the observers.

(6) We are facing an event that has never been seen in the history of mankind. We have shown that the entering of a comet in the atmosphere is distinctly different from the entering of an asteroid (a bare stone), and it is the most probable explanation of the Fatima event. Of the thousands of objects that have entered our atmosphere in the last centuries and of which we have records, all have been stones. We have not been able to find in the scientific literature one single case that could fit the characteristics of a comet as in this case. Thus:

We claim that Fatima is the only historical registered case of a comet collision with our planet. All others have been stones.

However, there might be an exception in case a comet with a rocky nucleus enters the atmosphere, vaporizes all the ices, and the rocky core gets to the ground.

(7) Chapter 7 shows that comet Fatima is associated with the meteor shower of the Psi Virginids. So, the Virgin Mary made a prediction using a meteor shower with her own name.

(8) We conclude that it was a small comet because the density of comets is about 1 gm/cm^3, and are mostly composed of dust, ices, water ice (H_2O), carbon monoxide (CO), and carbon dioxide (CO_2). CO_2 is used to cool ice cream, and it will evaporate at room temperature with no residue left. If the object had been a fragment of an asteroid, its density would have been 3 gm/cm^3 and would have left a residue after exploding, a rain of meteors that would have caused casualties. So this fact favors a comet and disfavors an asteroid fragment. Our facts concur with Chyba et al.[19]:

> *"The fate of bolides in the Tunguska size range is strongly dependent on the nature of the object. Had the Tunguska object been a 15 Mton **comet**, far less surface destruction would have resulted because of the much higher altitude of the airburst. **Comets incident with tens of kilotons of energy, explode so high in the atmosphere that they are scarcely noticed at the surface.**"*

(9) Since the event can be explained using a well-known physical phenomena, the event cannot be considered as a "miracle."

7. Limits of the scientific explanation

Dr. Chojnowski (personal communication), asked me a few numerical questions that I will try to answer to the best of my ability.

But since we are dealing with probabilities, we need first a reference value. Let us take as a reference Golf, and the odds of making a hole in one shot. This probability[43] is 1 in 12.500. It is a small probability, but nevertheless it happens all the time. My estimate is once per month.

1- "What are the odds of the comet falling but, unlike the cases of similar phenomena in Russia, no one being injured and nothing damaged in any way?"

- The probability depends on the size and composition of the object and the area where it falls, but since we lack this information it would be difficult to do that estimate.

2- "What would be the likelihood that the 10 year old uneducated girl would be able to persist in her claims, thereby bringing together 70,000 individual people to see an astrophysical event that actually happened?"

- That probability is different from zero and depended on her power of persuasion, which seems to be significant.

3- "What are the odds of the rain stopping and the clouds clearing so that the event could be seen at all?"

- After a storm passes, the sky will clear up. For the purposes of calculation, let us say that there is a storm 180 days per year (half of the year). Next let us assume 10 hours of daylight. Then the probability that the rain will stop at midday is 1 in 1800.

4- "What are the odds of a comet, rather than mere stones, falling to earth and creating the phenomena?"

- The NASA CNEOS database contains records of 856 fireballs observed between 1994 and 2021, an interval of 27 years. That means 32 fireballs per year, of sizes 1 to 30 meters, covering the range of estimated values of the Fatima event.

Let us assume that this rate was the same for the last 100 years, 1921-2021. Then the total number of events would have been about 3170, and all of them were stones, with no comet impacts.

Our coverage of celestial events is more incomplete the further we go back in time. If we go back 200 years (risky), the number would have been 6,400 events, all stones, no comets.

5- *"What are the odds of the comet coming at such an angle that it appears to be the Sun itself?"*

- The whole sky covers 3282 square degrees. Half that value is about 1641 square degrees seen in daylight. The diameter of the Sun is half a degree. Then the probability of the comet **"coming from the Sun's direction"** is 1 in 6,540.

The probability of **"falling on Cova da Iria"** is much smaller and can be calculated thus. The Miracle of the Sun was not seen from Leiria, 17.5 km away. Let us assume that the miracle was seen in a radius of 10 km. The radius of the Earth is 6371 km, and its cross-section is 127.5 million km². **The object could have fallen over any area of our planet**, but the probability of falling on Cova da Iria **"from any direction in the sky"** is 1 in 405,896.

Next, the probability of **"falling on Fatima on that date and at that time within 1 hour, coming from the Sun's direction,"** is 1 in 6540x405896x365x10 = 9,500,000,000,000. But we know the time of the event with an error of ±15 minutes, so the above number has to be multiplied by a factor of 2, or **a probability of 1 in 19,378,000,000,000**.

This number is the probability of a stone falling on Fatima. Above we calculated that in a hundred years, the probability of a comet falling is less than one in 3170. If we multiply the above number by 3170 we get one chance in 61,428,000,000,000,000. That is small!

6- *"What are the odds that a 10 year old uneducated Portuguese girl could correctly predict the happening of any astrophysical activity whatsoever?"*

- Nil. Zero. No way.

7- *"What would be the likelihood that such a girl would continue to predict, over the course of 3 months, any kind of astrophysical activity that would actually happen?"*

- Absolutely no way.

To make that kind of prediction, astronomers have to use an areas of Astronomy called "Astrometry" and "Celestial Mechanics", solve differential equations and do complex mathematical calculations.

We see that there is an unresolved issue, a conundrum that we cannot answer. The issue is the prediction itself. **The Lady predicted the date and the time of the miracle months in advance!** This cannot be explained scientifically. We can explain the dancing of the Sun, we can explain the falling of the Sun, we can explain the drying of the garments, but the prediction itself we cannot explain. It is beyond our current scientific comprehension and beyond the realm of Science.

So, incredibly, we are back to square one. From the point of view of Science, we can explain the event scientifically so it was not a miracle. However, using the definition of the Merriam-Webster dictionary, this prediction fulfills the definition of a miracle. Thus:

The miracle that was no miracle, was a miracle.

The analysis we made in the previous Sections is based on scientific facts backed up by scientific references. That is as far as we can go scientifically.

The event at Fatima was extraordinary. The probability of it happening by chance was

1 in 61.428.000.000.000.000

This is unique in the astronomical history of our planet. It interweaves Science and Religion in a way that to my knowledge no other reported miracle has ever been able to do. The scientific data was sitting there waiting to be interpreted.

The most outstanding result of this investigation is that, by validating the event but not validating the prediction, Science is actually validating the miracle. Science and Religion have never been so united.

Science Touched the Divine

Exhibit 21. The Battleship USS Iowa (BB-61) (The Big Stick), firing a full broadside of her nine 16"/50 and six 5"/38 guns during a target exercise near Vieques Island, Puerto Rico. Two blast waves are moving out of the tip of the barrels of the guns (arrowed). Whenever there is a large explosion, a blast wave follows that can potentially cause significant damage. It is believed that a blast wave caused some of the phenomena observed at Fatima. (Credit: Wikipedia Commons, National Archives and Records Administration, ARC 6396437, Phan J. Alan Elliot).

Research

CHAPTER 7
Comet Fatima: Analysis II

In Chapter 6, using 17 facts, we have concluded that a small comet caused the event that took place over Fatima. That was called "The Comet Fatima Hypothesis," CFH. In this Chapter, we will strengthen this hypothesis by doing additional mathematical calculations. If you have a hypothesis that can explain the 17 facts better than the CFH, please do not hesitate to write to me asap.

1- Basic Data of the Fatima Event

The analysis of the Fatima event did not finish with Chapter 6. It continues in this Chapter but becomes more technical. In Chapter 6, we examined 17 facts. We concluded that they were consistent with the disintegration of a small comet over the Earth's atmosphere. Here we do additional calculations and derive numerical parameters. These details are needed to build a scientific case. The first information we need are the parameters of the event. Using the information provided by this web site we retrieve some values:

https://www.suncalc.org/#/40.1789,-3.5156,3/2021.01.17/11:28/1/3

The time of the event and the event date are not critical for the results of this investigation. You can change the event by a few hours or even the date by a few days, and the results will be the same. These results appear in Table 1. With this information, we can go to the radiant meteor database of the International Astronomical Union, Meteor Data Center:

https://www.ta3.sk/IAUC22DB/MDC2007/Roje/rojelista.php?corobic_roje=0&sort_roje=0

Comet Fatima came from the Sun's direction, so our strategy is to search in the above Catalog those showers that took place in daylight. This information is shown in Table 2. We are looking for a shower with maximum activity near October 13. The result is almost at the end of Table 2. Shower 240 Psi Virginids has a maximum on October 11, only two days earlier. The previous shower was 16 days before, and the following shower was 11 days after. The initial identification of the 240 Psi Virginids shower, was made by Peter Veres, an orbit calculation expert, who works at the Minor Planet Center. So there is no doubt about the identification.

Table 1. Basic Parameters of the Fatima Fall

DATE, TIME, PLACE
DATE OF EVENT:	October 13, 1917
TIME:	Local Time=12:15 ±15 min
	Universal Time =11:15 ±15 min
LOCATION:	Cova da Iria, Fatima
LATITUDE:	N 39° 37' 54"
LONGITUDE:	W 08° 40' 23"
ALTITUDE:	358 m above sea level

SUN'S POSITION:
RIGHT ASCENSION:	198.01° = 13h 12m 10s
DECLINATION:	-7°.7
ALTITUDE:	40°.3
AZIMUTH:	158°.3

SHOWER'S RADIANT
RIGHT ASCENSION:	193.7° = 12h 54m 36"
DECLINATION:	-9.6°
SHOWER'S MAXIMUM:	201° = October 11th
VELOCITY OF METEORS:	21.1 km/s

2- Identification of the asteroid associated with the shower

Next, we would like to identify the parent body, if there is any. We have to compare the a = semi-major axis and q = perihelion distance values. Those are listed in two databases: The *"Comet Catalog of Orbits"* and the *"Potentially Hazardous Asteroids,"* freely available on the internet.

Table 2. 44 Daylight showers from the IAU Meteor Database

IAU # of Shower	Name of Shower	Sun Longit Show Max	Date of Maximum
115 DCS	Daytime Capricornids-Sagitt	301-325.1°	JAN 16-FEB 9
100 XSA	Daytime xi Sagittariids	304.9°	JAN 23
114 DXC	Daytime chi Capricornids	311.3°	JAN 30
116 DEQ	Daytime epsilon Aquariids	315.1°	JAN 30
117 DCQ	Daytime c Aquariids	324.7°	FEB 2
129 QPE	Daytime q Pegasids	354.5°	MAR 11
128 MKA	Daytime kappa Aquariids	359.7°	MAR 16
141 DCP	Daytime chi Piscids	19.2°	APR 9
144 APS	Daytime April Piscids	26-30.3°	APR 16-21
143 LPE	Daytime lambda Pegasids	29.7°	APR 20
419 DAC	Daytime April Cetids	29.7°	APR 20
153 OCE	Daytime S. omega Cetids	45.0-48.6°	MAY 5-9
152 NOC	Daytime N. omega Cetids	45.5-64.4°	MAY 6-25
351 DTR	Daytime Triangulids	46.0°	MAY 7
154 DEA	Daytime epsilon Arietids	48.1°	MAY 9
156 SMA	Daytime S. May Arietids	52.7°	MAY 13
354 DDT	Daytime delta Triangulids	53.0°	MAY 14
355 XIC	Daytime xi Cetids	54.0°	MAY 15
294 DMA	Daytime May Arietid Cmplx	55°	MAY 16
155 NMA	Daytime N. May Arietids	55.0°	MAY 16
171 ARI	Daytime Arietids	76.7°	JUN 7
172 ZPE	Daytime zeta Perseids	78.6°	JUN 9
325 DLT	Daytime lambda Taurids	85.5°	JUN 15
173 BTA	Daytime beta Taurids	96.7°	JUN 28
174 TAS	Daytime theta Aurigids	96°	JUN 28
185 DBA	Daytime beta Andromedids	100.5°	JUL 1
188 XRI	Daytime xi Orionids	117.7-121.9°	JUL 18-22
189 DMC	Daytime mu Cancrids	126°	JUL 27
377 DMO	Daytime Monocerotids	137.0°	AUG 7
202 ZCA	Daytime zeta Cancrids	147°	AUG 15
203 GLE	Daytime gamma Leonids	148.7°	AUG 18
204 DXL	Daytime chi Leonids	154°	AUG 22
426 DCR	Daytime Craterids	158.0°	AUG 28
381 DPL	Daytime pi Leonids	174.0°	SEP 13
212 KLE	Daytime kappa Leonids	180.7-184.2°	SEP 20-23
222 DDl	Daytime delta Leonids	183°	SEP 22
223 GVI	Daytime gamma Virginids	184°	SEP 23
221 DSX	Daytime Sextantids	186.1°	SEP 25
240 DFV	Daytime Psi Virginids	202°	OCT 11
232 BCN	Daytime beta Cancrids	213-214°	OCT 22-23
251 IVI	Daytime iota Virginids	223°	NOV 1
261 DDC	Daytime delta Scorpiids	254°	DEC 3
S262 KLI	Daytime kappa Librids	259°	DEC 7

00240	DFV			Daytime psi Virginids				
Activity	S. Lon	RA [deg] J2000	DE	dRA	dDE	VG [km/s]	a [AU]	q [AU]
annual	202	193.7	-9.6			21.1	1.513	0.525
Parent body:								
Notes:								

Exhibit 22. The Meteor Data Center page[46], where the identification of the shower is made. We can calculate the eccentricity of the orbit, and we find e = 0.653. RA of radiant: 193.7° = 12h 54 m 36s, DEC of radiant = -9° 36," velocity = 21.1 km/s, rather fast. (IAU Meteor Data Center).

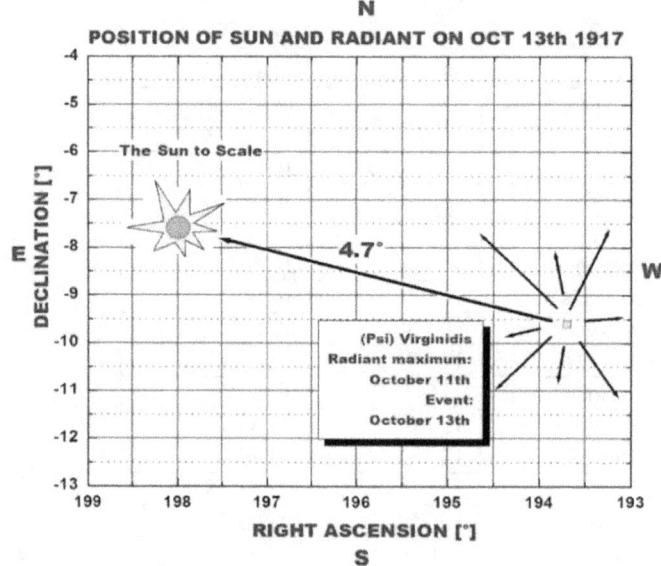

Exhibit 23. The Psi Virginids meteor shower is shown in the sky with the Sun plotted to scale on the event's date. The shower radiant emits meteors in all directions, and one of them came precisely in the direction of the Sun, producing the Miracle of Fatima. Suppose we adopt an error of 1/4 of the Sun's diameter. In that case, the probability of this meteor going in front of the Sun is about 1/118, a small probability but not that small. (IF).

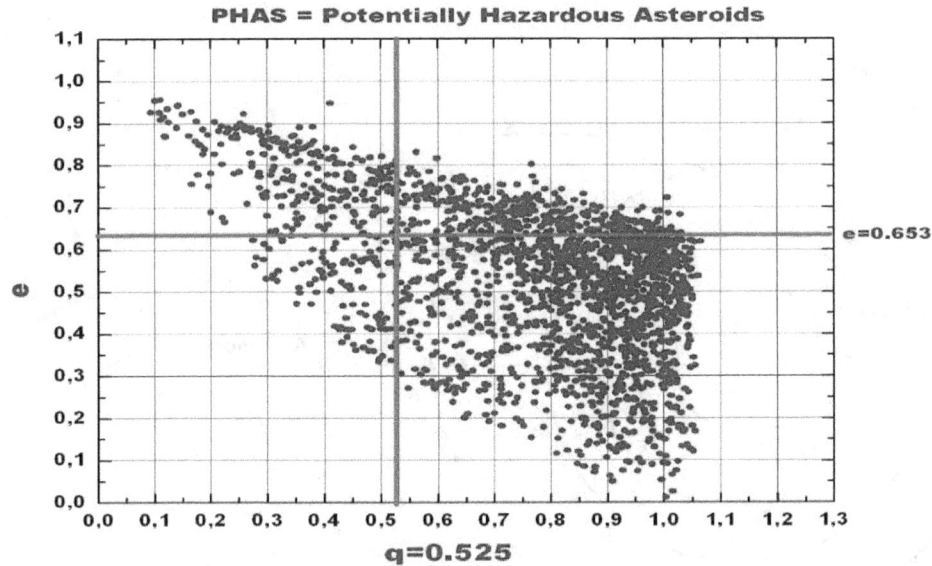

Exhibit 24. The parameters of the ψ (psi) Virginids shower, placed over the distribution of PHAS. We need to enlarge the plot to find the parent body. There are 2728 PHAs as of June 2020. (IF).

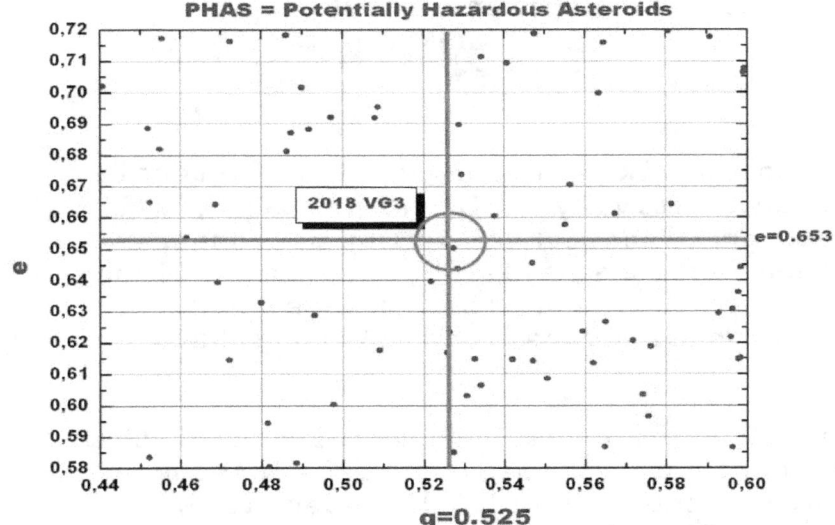

Exhibit 25. Here is the plot showing the distribution of Potential Hazardous Asteroids (PHAs) in the e versus a space. Asteroid 2018 VG3 is the parent body of the ψ (Psi) Virginids shower. The comet catalog did not show any match. (IF).

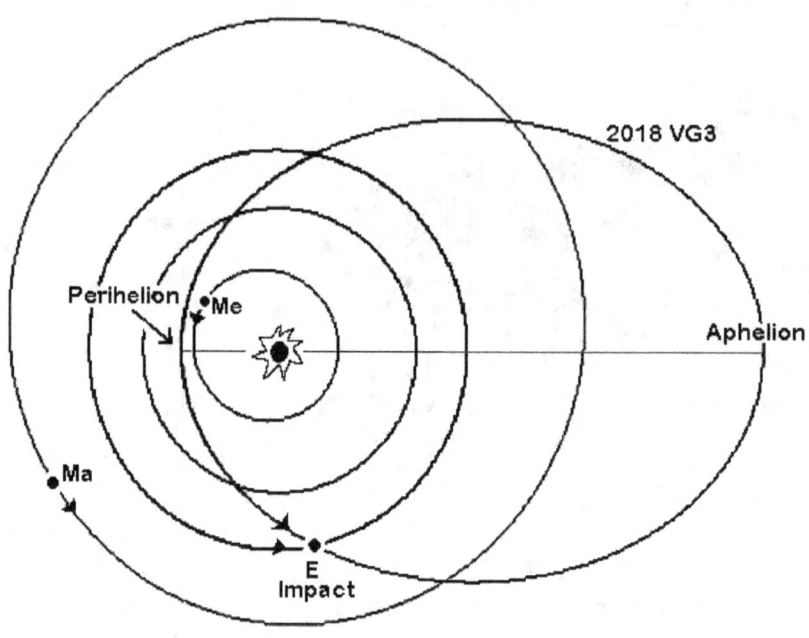

Exhibit 26. The orbit of asteroid 2018 VG3, proposed parent of the Fatima Comet, and its meeting with Earth on 1917 October 13, in the dayside. The orbit does not reach to Jupiter. (IF).

Next, we would like to know if this asteroid exhibits cometary activity. To do that, we will create the Secular Light Curve, using astrometric-photometric observations available in the Minor Planet Center site. These results are shown in Exhibit 27 (the phase curve) and 28 (the Secular Light Curve).

From Exhibit 28, an absolute magnitude of the object can be derived: V(1,1,0) = 20.0. Assuming an albedo = 0.04, the object's diameter is 670 m (2233 feet) approximately.

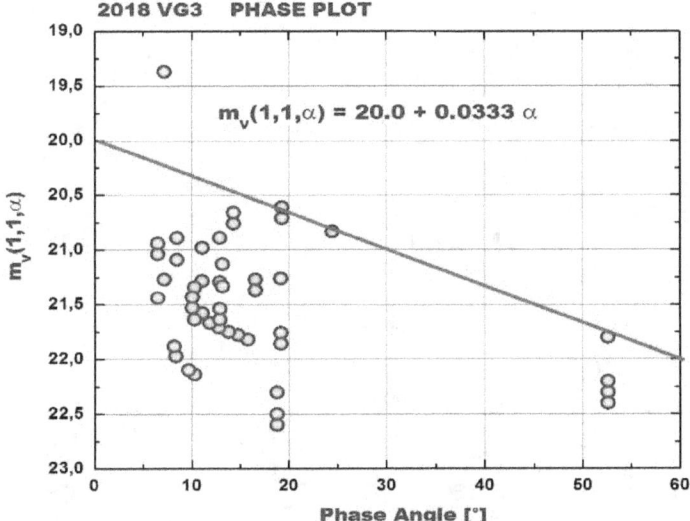

Exhibit 27. The phase curve of parent body 2018 VG3 has a phase coefficient β = 0.033±0.006, in agreement with the Taurid Complex[28] members that have a mean value < β > = 0.036±0.006. (IF).

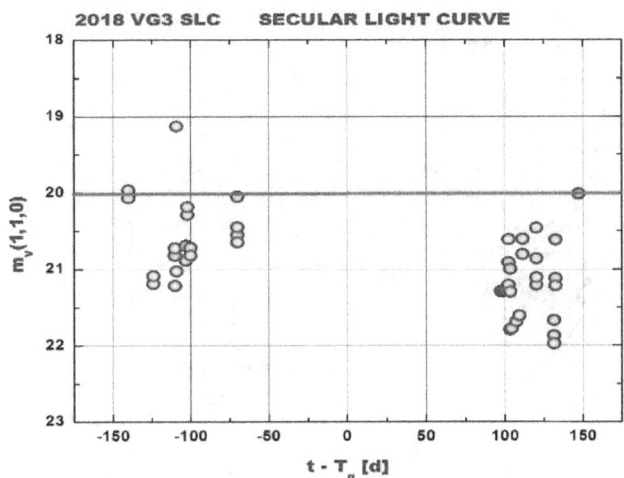

Exhibit 28. The secular Light Curve of 2018 VG3 does not have enough data points to conclude if this parent body has cometary activity or not. In the x-axis, the time with respect to perihelion is given. In the vertical axis, the absolute magnitude is plotted. The plot is flat except for one single data point. An active comet is expected to show a bump or series of data points well above the horizontal line. (IF).

Table 3. Parent Asteroid 2018 VG3
a = semi-major axis = 1.508 AU
e = eccentricity = 0.65
q = perihelion distance = 0.527 AU
i = 6.0°
P = orbital period = 1.85 years
Q = Aphelion Distance = 2.489 AU
ϖ = orbital longitude = 280°
m(1,1,0) = H = absolute magnitude = 20.0
D = diameter ~670 meters
β = phase coefficient = 0.033±0.006

3- Does the Fatima comet belong to the Taurid Complex?

In Table 3, we have derived an eccentricity of the orbit of 0.65. We also have the semi-major axis, a=1.508. With this information, we can plot our Fatima comet over the Taurid Complex distribution, Exhibits 29 and 30.

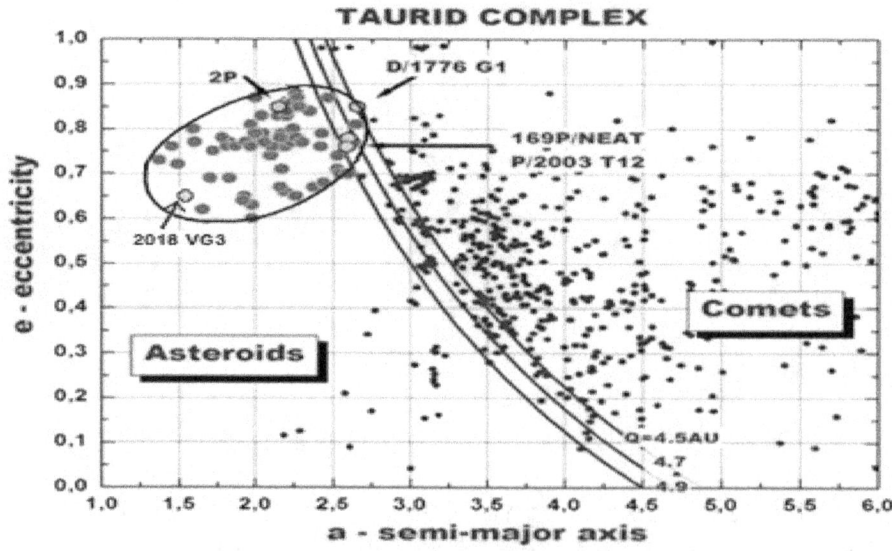

Exhibit 29. Members of the TC plotted on the **eccentricity** versus **semi-major axis** diagram lie on an ellipsoidal area. Parent asteroid 2018 VG3 lies completely inside the distribution, suggesting that it might belong to the Complex. (Credit: Vincenzo Orofino and this author[28]).

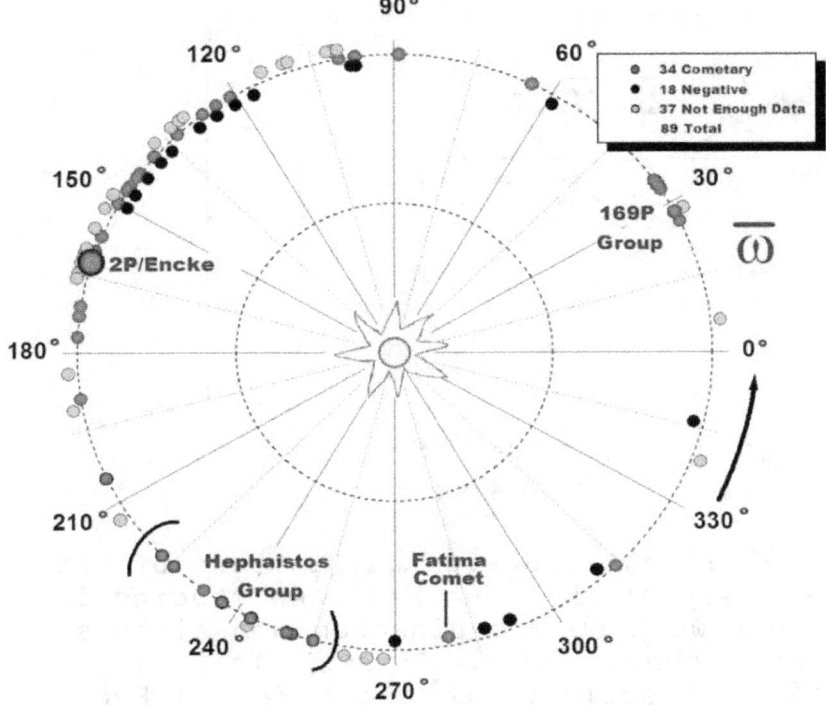

Exhibit 30. The location of the parent body 2018 VG3 in the polar diagram of the Taurid Complex. The variable ϖ measures the position along the orbit. The orbit is elliptical, but for clarity, we have drawn it as a circle. Although the object lies far from 2P, it still has other members of the Complex around. (Credit: Vincenzo Orofino[37] and IF).

<> <> <> <> <>

We conclude that it lies well inside the TC ellipse, and thus there is a strong suspicion that it may belong to the TC.

In Exhibit 30, the data points have been drawn at slightly different distances from the center to help visualize the three different types of objects: active, inactive, and with not enough data to decide.

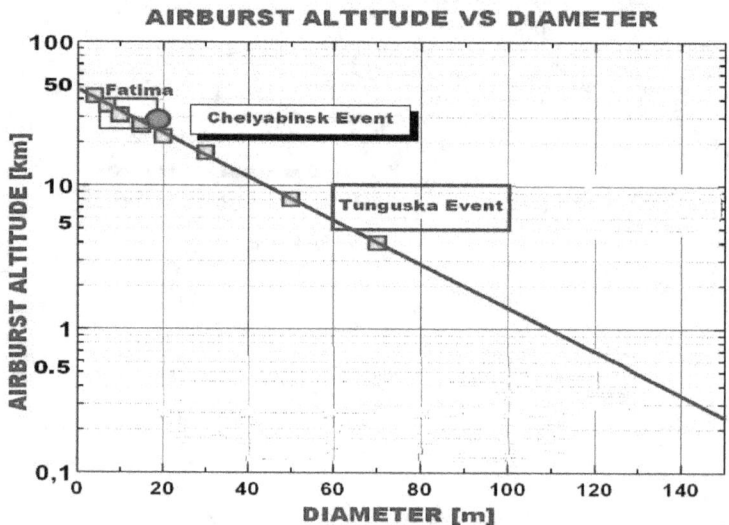

Exhibit 31. Burst altitude versus diameter of the object. The Chelyabinsk event has been studied in detail[38,42]. Thus we know its diameter and airburst altitude that seems to validate Tricarico's calibration[39]. We estimate that the Fatima Event was not as strong as the Chelyabinsk Event. Thus, its diameter can be constrained as 10 < D < 20 meters, and its burst altitude 25 < altitude < 40 km. (IF).

4- The Impact Effect Program

Robert Marcus, H. Jay Melosh and Gareth Collins[44] of The Imperial College of London and Purdue University have written a program to calculate the Earth Impact Effects of an asteroid or comet. The calculation is a guide on what may happen rather than a precise result.

We would like to use this program to calculate some values of the Fatima comet impact. In Exhibit 32, we have determined some limits, 10 < diameter < 20 meters, and 25 < burst altitude < 40 km. Since the Fatima body was a comet, we adopt water density, 1000 kg/m³. From the event's parameters listed in Table 1, we adopt the inclination angle as 40°. Inserting these values into the program, we find the results of Exhibit 32.

Your Inputs:

Distance from Impact: **25.00 km (= 15.50 miles)**
Projectile diameter: **20.00 meters (= 65.60 feet)**
Projectile Density: **1000 kg/m^3**
Impact Velocity: **21.00 km per second (= 13.00 miles per second)**
Impact Angle: **40 degrees**
Target Density: **2500 kg/m^3**
Target Type: **Sedimentary Rock**

Energy:

Energy before atmospheric entry: **9.24 x 10^{14} Joules = 220.65 KiloTons TNT**
The average interval between impacts of this size somewhere on Earth is **34.4 years**

Major Global Changes:

The Earth is not strongly disturbed by the impact and loses negligible mass.
The impact does not make a noticeable change in the tilt of Earth's axis (< 5 hundreths of a degree).
The impact does not shift the Earth's orbit noticeably.

Atmospheric Entry:

The projectile begins to breakup at an altitude of **84000 meters = 276000 ft**
The projectile bursts into a cloud of fragments at an altitude of **32700 meters = 107000 ft**
The residual velocity of the projectile fragments after the burst is **14.3 km/s = 8.9 miles/s**
The energy of the airburst is **4.94 x 10^{14} Joules = 0.12 x 10^0 MegaTons**.
No crater is formed, although large fragments may strike the surface.

Air Blast:

What does this mean?

The air blast will arrive approximately **2.08 minutes** after impact.
Peak Overpressure: **683 - 1370 Pa = 0.00683 - 0.0137 bars = 0.0969 - 0.194 psi**
Max wind velocity: **1.6 m/s = 3.59 mph**
Sound Intensity: **57 dB (Loud as heavy traffic)**

Exhibit 32. Results of the Impact Effects Program. The impact has the Destructive Power of 8 Hiros, a blast wind velocity of 5.8 km/h, and the air blast will arrive 2.08 minutes after impact. The Impact Effect Program calculated an energy of 0.12 Megatons. This is equivalent to ~8 Hiros, the estimated DP of the Fatima Comet (Exhibit 33). (Credit: Collins, Melosh and Marcus[44]).

5- Destructive power of the Fatima comet

We would like to estimate the Destructive Power (DP) of the fall. **We will define a new measuring unit, 1 Hiro**(shima), the destructive power of the atomic bomb dropped over the Japanese city of the same name (Exhibit 33).

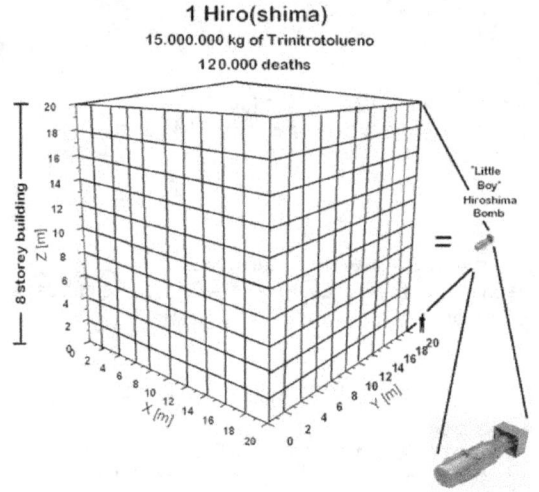

Exhibit 33. Definition of 1 Hiro, the destructive power of the Hiroshima Bomb. It is equivalent to the size of a building of 20x20x20 meters made up of Trinitrotolueno. The bomb was 3 meters (9 feet) in length. (IF).

6- Analysis of the Kamchatka Event

In Exhibit 34, we show an image of the Kamchatka event imaged from a NASA satellite on December 18, 2018. We can see the trail's shadow, and it exhibits a resolved width. We will calculate the width to discern if this trail could cause a solar eclipse. The result is positive. The trail has a Half-Width Half Maximum of 7 km. At a distance of 35 km, this represents 11° in the sky, while the Sun's diameter is 0.5°.

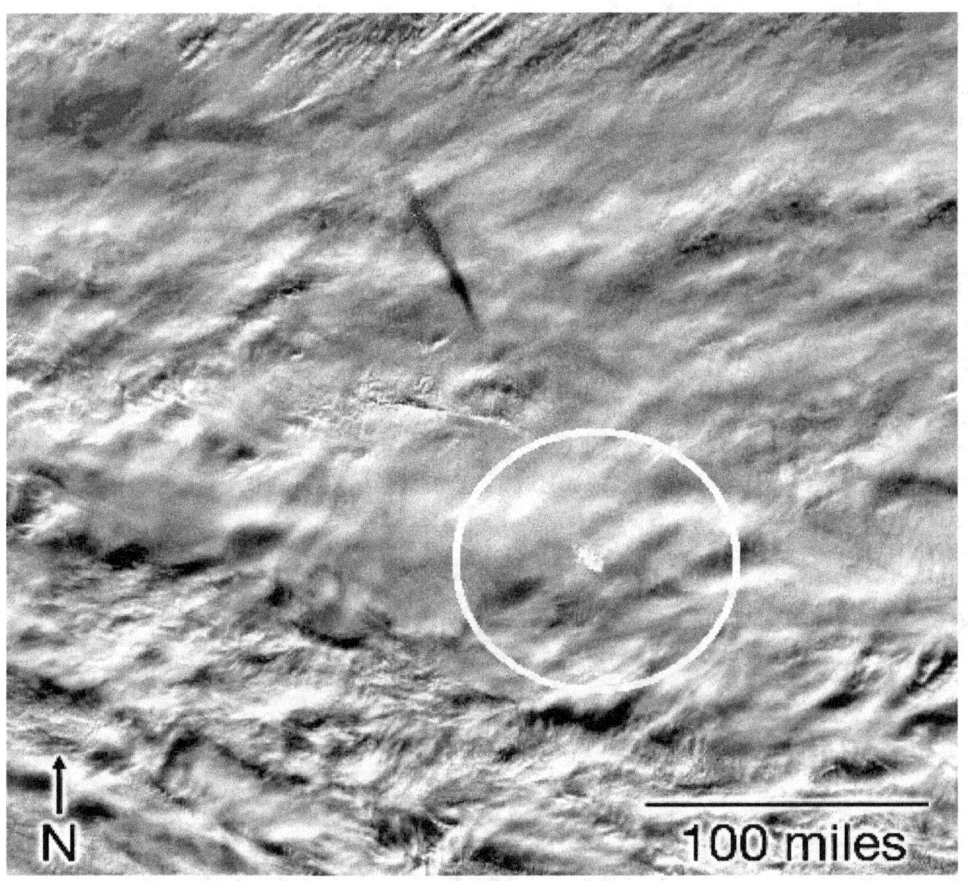

Exhibit 34. The Multi-Angle Spectrometer-Radiometer (MISR), aboard NASA's Terra Satellite, observed the Kamchatka event. The yellow spot at the center is the fireball trail after it was super-heated in the air as it passed through. Above the center-left, the shadow of the meteor's trail is visible, cast over the cloud tops, and elongated due to the low Sun angle. (Credit: NASA Terra Satellite).

Since the Sun covers only 0.5°, there is plenty of space to produce a solar eclipse, proving that the solar dimming proposed for the Fatima miracle is physically possible and depends solely on the object's mass. The KAM event was an asteroid. Had it been a comet, the cloud's size could have been much more significant because of the material's amount evaporated.

- EPILOGUE

What an incredible journey. I am as surprised as you are. I was not trying to debunk the Fatima Miracle. I was just trying to explain it, and I was not expecting this result.

To me one of the most significant facts was Figure 33 showing 21 people looking at the Sun at the same time with no eye protection. Of the 70,000 people attending, not even one reported blindness.

However, there are two issues pending. First, in the normal blast waves there is a strong sound that accompanies the wave. But after reading more than 1600 pages, the sound was not reported by anybody. There are two possibilities. The rain stopped falling before midday. While raining, there must have been lightning and the rumble of thunder which is very similar to the sound of the blast wave. Maybe the blast wave sound was there but it was confused with the rumble of thunder and thus nobody reported it.

There is another alternative. A comet has a mean density of about ~0.6 gm/cm^3 thus it is very fragile. When it enters our planet's atmosphere it disintegrates vaporizing itself slowly and gradually, while in the case of a stone rapidly and suddenly. Perhaps there was no sound or the sound was more muted and longer lasting than in the case of a stone. This issue could be resolved if there were evidence of barometric fluctuations. My efforts to find such records from 1917 have failed up to now.

Second, Fact #17 was evidence that some residue may have reached the ground. After 104 years have passed, it is practically impossible to find that residue because of rain carrying it away. But perhaps it could be found in tree rings. When tree rings were examined in Tunguska, it was found that the Tunguska's cosmic body had impregnated the ground with a substance that increased the width of the tree rings.

This investigation will continue.

INDEX OF CITATIONS AND COLLABORATORS

- Bordiga, F., 21
- Campbell, Steuard, 94, 95
+ Chamberlin, Alan, B., 20
+ Chyba, C., 131, 159
- **Chojnowski, Peter 94, 100, 103, 108, 131, 158, 165**
+ Collins, G. S., 147, 148, 162
- De Almeida, Avelino, 47, 86, 100, 157
- De Marchi, John, 47, 54, 75, 98, 103, 106, 107
- Denning, W.F., 28
+ Elst, Eric, 23, 24, 26, 162
+ **Fedorova, Natalia, 38, 165**
- Galileo Galilei, 13, 15
- Garret, Almeida, 75, 82, 101, 103, 107, 109, 110, 112, 115
- Gens, Pereira, 126
+ **Gracia Montoya, Alvaro, 32, 165**
- Haffert, John, M., 117, 123, 158
- Harrington, R. G., 23, 24, 25, 26, 27, 159
- Helin, Eleanor, 24, 25
- Herschel, William, 20
+ **Horálek, Petr, 104, 165**
- Jaki, Stanley, 91
- Jefferson, Thomas, 19
- Kelvin, Lord, 13
- Kosolapov, M., 127
- Kulik, L. A., 33, 37, 38
- Lourenco, Inacio, 84, 120

- Madigan, Leo, 93
+ Marcus, Robert, 147, 148, 162
- Marsden, Brian, 25
- **Michael Mattiazzo, 52, 165**
- McClure, Kevin, 94, 158
- Meessen, Auguste, 94, 95, 159
+ Melosh, H., 147, 148, 162
- Napier, W., 29, 160
- Nickell, Joe, 95
- **Orofino, Vincenzo, 29, 30, 31, 145, 146, 160, 165**
- Piazzi, Gioacchino, 20, 21, 22
- Pinto, Domingo, 84
+ Pizarro, Guido, 24, 26, 162
- Radford, Benjamin, 92
- Reis, Dominic, 123
- Rubtsov, V., 117, 158
- Sagan, Carl, 96
- Shrbený, L, 115, 161
- Silberman, Joseph, 121, 122
- Simons, Paul, 95
- Spenak, Fred, 50, 51, 157
- Spurnný, P., 115, 161
- Tricarico, Pascuale, 147, 161
+ **Veres, Peter, 138, 155, 165**
- **Voltmer, Sebastian, 118, 165**
- **Ward, Bill, 118, 165**
- Wilson, A. G., 23, 24, 25, 26, 28, 159

- Glossary

- **Absolute magnitude of an asteroid, "H" or m(1,1,0).** The brightness would have a comet at a distance of 1 AU from both the Sun and Earth, and zero phase angle.

- **Asteroid.** In this work, we adopted the definition of **"asteroid"** given by A.E. Rubin & J.N. Grossmann (Meteoritics & Planetary Science 45, 114-122, 2010), which incorporates the recommendations of the IUA Commission F1 (https://www.imo.net/definitions-of-terms-in-meteor-astronomy-iau/). According to this definition, any body with diameter > = 1 m is classified as an asteroid. So *the objects of Tunguska and Chelyabinsk are not meteoroids but asteroids.* Although most asteroids are constrained between the orbits of Mars and Jupiter in the so called The Main Belt, some stray beyond those limits and approach planet Earth, with which they may collide occasionally. These are the PHAs. My thanks to Vincenzo Orofino for this clarification.

- **Aphelion.** Objects inside the Solar System follow elliptical orbits. The place where the object is furthest from the Sun is known as the aphelion.

- **Astronomical Unit, (AU).** A measurement of distance in Astronomy is equivalent to the Sun's distance to theEarth, 149.600.000 km or 93.000.000 miles.

- **Comet.** A celestial object containing dust and ices. When they sublimate (evaporate) near the Sun, they exhibit a tail and a coma (atmosphere). Comets have been called "dirty ice balls."

- **Destructive Power.** The destructive capability of a comet or an asteroid is measured in Hiros.

- **Encke's Comet.** 2P/Encke is the second periodic comet numbered after 1P/Halley. It is an odd object because of several properties, like the fact that it has more than 30 meteor showers associated. Its importance is that it is the parent of the Taurid Complex.

- **Fireball.** *It* is defined as "a large bright meteor," so although the Chelyabinsk Cosmic Body is correctly listed in the CNEOS database as a fireball, it is actually an asteroid.

- **1 Hiro.** It is equivalent to the Destructive Power of the bomb of Hiroshima.

- **Magnitude.** The brightness of an astronomical object is measured on an inverted scale. The star Vega = Alpha Lyra has magnitude 0, while the faintest star that can be seen with the naked eye on a clear night is 6. Large telescopes can reach magnitude 20 and beyond.

- **Meteoroid.** Refers to the body of small object meters in diameter, going around the Sun before entering the atmosphere.

- **Meteor.** The meteoroid after it has entered the atmosphere.

- **Meteorite.** The meteoroid on the ground.

- **Perihelion.** Objects inside the Solar System follow elliptical orbits. The place where the object is nearest to the Sun is known as the perihelion.

- **Potentially Hazardous Asteroid, PHA.** It is defined as an object with an $H<22$ and $MOID < 0.05$ AU = 1.54 Earth radii. I owe this definition to Peter Veres.

- **Orbital Elements.** Comets and asteroids go around the Sun in elliptical orbits which are defined by several physical parameters: a = semi-major axis, e = eccentricity, P =

orbital period, i = inclination of the orbit, w = longitude of perihelion and Ω = longitude of ascending node. Tq = time of perihelion.

- **Semi-major axis, a.** Asteroids have elliptical orbits around the Sun. The largest axis of the ellipse is called the semi-major axis, a.

- **Taurid Complex, TC.** It is a group of more than 100 objects that have disintegrated from the parent comet 2P/Encke, and that follow it in its orbit. The TC includes comets, asteroids, pebbles, debris and lots of dust.

- REFERENCES

+1 "Earth Impact Database." University of New Brunswick.
http://www.passc.net/EarthImpactDatabase/New%20website_05-2018/Index.html

+2 In FATIMA XXI. Duarte, Marco Daniel. "As cores e as formas do Sol: O Milagre do Sol de 13 Outubro de 1917, no corpus artistico da Cova da Iria." P. 185-186. Editor: Carlos Cabecillas. Fatima Santuary.

+3 "Catalog of Earth Impact Structures," Siberian Center for Global Catastrophes, Russian Academy of Sciences.
https://web.archive.org/web/20090716232605/http://omzg.sscc.ru/impact/index1.html

+4 De Marchi, J., (1947). "The True Story of Fatima."
file:///F:/0%20RESPALDO%20ACTUAL/000001%20TUNGUSKA%20NUEVO%2019111/0001%20PAPERS%20POR%20LEER%20200127/The-True-Sto

+5 Church Editor, 2017. "Is this a Photo of Fatima's Miracle of the Sun?."
https://www.churchpop.com/2017/05/11/is-this-a-photo-of-fatimas-miracle-of-the-sun-the-truth-behind-the-popular-photo/

+6 Spenak, Fred. "Canon of Solar Eclipses."
https://eclipse.gsfc.nasa.gov/SEcat5/SE1901-2000.html

+7 de Almeida, Avelino. "O Milagre de Fatima."
file:///F:/0%20RESPALDO%20ACTUAL/000001%20TUNGUSKA%20NUEVO%2019111/0001%20PAPERS%20POR%20LEER%20200127/30722706-The-Miracle-of-Fatima-El-Milagro-de-Fatima-O-Milagre-De-Fatima.pdf

+8 Ferrín, I., 2021. "Next Asteroid Impact." In print.

+9 McClure, K., 1983. "The Evidence for the Visions of the Virgin Mary."
https://archive.org/details/evidenceforvisio00mccl/mode/2up

+10 Wikipedia, "Solar Eclipse of April 17, 1912."
https://en.wikipedia.org/wiki/Solar_eclipse_of_April_17,_1912

+11 Jubier, X. "Hybrid Solar Eclipse of 1912 April 17 in Europe."
https://web.archive.org/web/20100709161823/http://xjubier.free.fr/en/site_pages/solar_eclipses/HSE_1912O417_pg01.html

+12 Chojnowski, P., 2010. "Miracle of the Sun."
https://web.archive.org/web/20170829041305/http://www.fatimacrusader.com/cr96/cr96pg79.pdf

+13 Chojnowski, P., "Fatima Challenge Conference," Rome, Italy, May 3, 2010.
https://www.youtube.com/watch?v=WN-61UJN4aM

+14 In FATIMA XXI. Antonio Marujo. "Apenas afirmar o que vi o 13 de Outubro de 1917 em testemunhos e na imprensa." P. 128. Editor: Carlos Cabecillas. Fatima Santuary.

+15 Rubtsov, V., 2013. "Reconstruction of the Tunguska Event of 1908: Neither an Asteroid nor a Comet Core."
https://arxiv.org/ftp/arxiv/papers/1302/1302.6273.pdf

+16 Haffert, John, 1961. "Meet the Witness of the Miracle of the Sun." The American Society for the Defense of Tradition, Family and Property - TFP. Spring Grove, Penn., 17362. Pages 7-12.
https://web.archive.org/web/20161128200651/http://johnhaffert.org/wp-content/uploads/2013/08/Meet-the-Witnesses-91511.pdf#page=90

+17 Wikipedia, "Blast Waves."
https://en.wikipedia.org/wiki/Blast_wave

+18 "Our Lady of Fatima," Roman Catholic Church of Our Lady of Fatima, White City, London, UK.
https://parish.rcdow.org.uk/whitecity/our-lady-of-fatima/

+19 Chyba, C., P. Thomas, and K. Zahnle 1993. "The 1908 Tunguska Explosion: Atmospheric Disruption of a Stony Asteroid." Nature 361, p. 40-44.
http://www.igpp.ucla.edu/public/mkivelso/refs/PUBLICATIONS/Chyba%20Tunguska.pdf

+20 Wikipedia. "Fatima Miracle."
https://en.wikipedia.org/wiki/Miracle_of_the_Sun

+21 Meessen, Auguste, 2005. "Apparitions and Miracles of the Sun." International Forum in Porto: "Science, Religion and Conscience," October 23-25, 2003.
https://www.meessen.net/AMeessen/Apparitions_and_Miracles_of_the_Sun.pdf

+22 Alvarez, L. W. et al. (1980). "Extraterrestrial Cause for the Cretaceous-Tertiary Extinction." Science, 208, 1095.

+23 Alvarez, L. W., (1983). "Experimental evidence that an asteroid impact led to the extinction of many species 65 million years ago." PNAS, 80, 627-642.

+24 Ferrín et al., 2012. "The 2009 Apparition of Methuselah comet 107P/Wilson-Harrington: A case of comet Rejuvenation?"
https://arxiv.org/ftp/arxiv/papers/1205/1205.6874.pdf

+25 Ferrín, et al., 2016. "Secular light curves of 25 members of the Themis family of asteroids, suspected of low-level cometary activity."
https://arxiv.org/ftp/arxiv/papers/1703/1703.09889.pdf

+26 Ferrín, et al., 2018. "Secular light curves of NEAS 2201 Oljato, 3200 Phaeton, 99943 Apophis, 162173 Ryugu, 495848 and 6063 Jason."
https://arxiv.org/ftp/arxiv/papers/1807/1807.11157.pdf

+27 Ferrín, et al., 2019. "Secular and rotational light curves of 6478 Gault."
https://arxiv.org/ftp/arxiv/papers/1906/1906.10195.pdf

+28 Ferrín, I., and Orofino, V., 2021. "Taurid Complex Smoking Gun: Detection of cometary activity."
https://arxiv.org/ftp/arxiv/papers/2011/2011.13078.pdf

+29 Denning W. F. 1928. J. Br. Astron. Assoc., 38, 302.

+30 Napier, W., 2010. "Paleolithic extinctions and the Taurid Complex."
https://academic.oup.com/mnras/article/405/3/1901/966774

+31 Krinov, E. L., 2009. "The Tunguska Meteorite," International Geology Review, 2, 1960, Issue 1.
https://www.tandfonline.com/doi/abs/10.1080/00206816009473542

+32 "Tunguska Event - Siberia 1908."
http://earthsci.org/space/space/tunguska/tunguska.html

+33 Wikipedia, 2020. "Tunguska Event."
https://en.wikipedia.org/wiki/Tunguska_event

+34 Kresak, L., 1978. The Tunguska object: A fragment of comet Encke? Bull. Astron. Inst. Czech., 29, 129-134.

+35 Zotkin, I., 1966. Anomalous twilight associated with the Tunguska Meteorite. Meteoritica, 27, 109.

+36 Andreev, G. V., 1990. Was the Tunguska 1908 event caused by an apollo asteroid? Asteroids, Comets, Meteor Conference, 489-492.

+37 Asher, D. J., Steel, D. I., 1998. On the possible relation between the Tunguska bolide and comet Encke. Planet. Space Science, 46, 205-211.

+38 Brown, P., et al., 2013. "A 500 kiloton airburst over Chelyabinsk and an enhanced hazard from small impactors." Nature, 503, 238-241.

+39 Tricarico, Pasquale, 2016. "The Near-Earth Asteroid Population from Two Decades of Observations."
https://arxiv.org/pdf/1604.06328.pdf

+40 Yegorov, Oleg. "The mysterious Tunguska explosion of 1908."
https://www.rbth.com/arts/2017/05/30/the-mysterious-tunguska-explosion-of-1908_773137

+41 Shrbený, L., Spurný, P., (2012). "Determination of Velocity of Fireballs from High-Resolution Light Curves." Asteroids, Comets, Meteors, 2012. Proceedings of the conference held May 16-20, 2012 in iigata, Japan.LPI Contribution No. 1667, id. 6437.

+42 Popova, O., et al., 2013. "Chelyabinsk Airburst, Damage, Assessment, Meteorite

Recovery and Characterization." Science, 342, 1069-1073.

+43 "The Odds of Making a Hole in One."
https://www.ahno.com/americanhno-blog/odds-of-making-a-hole-in-one#:~:text=The%20odds%20of%20making%20a%20hole%20in%20one%20are%2012%2C500,drop%20to%202%2C500%20to%201.

+44 Collins, G. S., Melosh, H. J., Marcus, R. A., (2005). "Earth Impact Effects Program: A Web-based computer program for calculating the regional environmental consequences of a meteoroid impact on Earth."
https://onlinelibrary.wiley.com/doi/epdf/10.1111/j.1945-5100.2005.tb00157.x

+45 Ferrín, I., 2006. Secular Light Curves of comets, II 133P/Elst-Pizarro, an asteroidal belt comet." Icarus, 185, 523-543.

+46 International Meteor Data Center,
https://www.ta3.sk/IAUC22DB/MDC2007/

- About the author

- Born in Spain.

- Ph. D. in Astro-Geophysics, University of Colorado, Boulder, Colorado, USA.

- Full Professor
  Institute of Physics,
  University of Antioquia
  Solar, Earth & Planetary Sciences (SEAP) Group, Medellín, Colombia

- Full Professor
  University of the Andes
  Department of Physics
  Mérida, Venezuela

- Discoverer of 12 new minor planets inside the Solar System:

  38628, 127870, 149528, 161278, 366272 Medellín, 347940, 159776, 196476, 201497, 128166, 423624, 189310

- Has identified a new category of comets called "Lazarus Comets."

https://www.youtube.com/watch?v=DfQoWFXzaPk

- Please send your feedback at

  nextimpact2022@gmail.com

- If you would like to receive the monthly newsletter on asteroid impacts send "subscribe" to the email

  liadanewsletter2021@gmail.com

- If you liked this book please leave a review at Amazon.com

Figure 49. At the Tesoro shopping mall in Medellin. Decades ago, Medellin had a bad reputation due to drug dealings. Today it is a modern, exuberant city that has won the *"Best City of the World"* award in 2008. Known for its food, flowers, and beautiful women, it is called the *"the city of eternal spring."*

Figure 50. Observing with the 1 meter Schmidt telescope of the National Observatory Of Venezuela, at 3600 meters of altitude above sea level, in Merida. When the sky Is clear the atmosphere is incredibly transparent and dark.

- Acknowledgements

- **Virginia Rathbun**, for her careful correction of the manuscript, advice and suggestions.

- **Dr. Peter Chojnowski**, for his many suggestions to improve the book, his questions, concepts and ideas.

- **Dr. Peter Veres**, of the IAU Minor Planet Center, for sending me in the right direction when I was lost.

- **Dr. Brad Schaefer**, of Louisiana State University, for his many suggestions that improved the scientific quality of the manuscript.

- **Dr. Sabatino Sofia**, of Yale University for checking the calculation of the probability of a random event for Fatima.

- **Dr. Vincenzo Orofino**, of the University Of Salento, Lecce, Italy, for his teachings and collaboration.

- **Dr. Natalia Fedorova**, for his permission to reproduce a painting by Nicolay Fedorov.

- **Dr. Bill Ward**, for his spectrum of a meteor.

- **Dr. Sebastian Voltmer**, for his spectrum of a comet.

- **Artist Alvaro Gracia Montoya**, a motion graphic and 3D designer of Birmingham, creator of MetaBallStudio, for his permission to reproduce his art.

- **Peter Horálek**, for his permission to reproduce his image of the meteor shower of the Perseids.

- **Michael Mattiazzo**, of Australia, for his image of comet 2P/Encke.

www.ingramcontent.com/pod-product-compliance
Lightning Source LLC
Chambersburg PA
CBHW080456220526
45465CB00006B/2286